Conversations About Bernstein

DAVID DIAMOND

CHRISTA LUDWIG

MSTISLAV ROSTROPOVICH

PAUL MYERS

HAROLD SCHONBERG

JOHN MAUCERI

LUKAS FOSS

JUSTIN BROWN

FREDERICA VON STADE

JOAN PEYSER

JERRY HADLEY

JONATHAN MILLER

CAROL LAWRENCE

MEMBERS OF THE NEW YORK,

VIENNA, AND

ISRAEL PHILHARMONICS

Conversations About Bernstein

Edited and with an Introduction by

WILLIAM WESTBROOK BURTON

New York Oxford

OXFORD UNIVERSITY PRESS 1995

OXFORD UNIVERSITY PRESS

Oxford New York Toronto
Delhi Bombay Calcutta Madras Karachi
Kuala Lumpur Singapore Hong Kong Tokyo
Nairobi Dar es Salaam Cape Town
Melbourne Auckland Madrid

and associated companies in
Berlin Ibadan

Published by Oxford University Press, Inc.,
200 Madison Avenue, New York, New York 10016

Library of Congress Cataloging-in-Publication Data
Conversations about Bernstein / edited and with an introduction by
William Westbrook Burton.
p. cm. Includes index.
ISBN 0-19-507947-7
1. Bernstein, Leonard, 1918–1990.
2. Musicians–Interviews.
I. Burton, William Westbrook.
ML410.B566C66 1995 780'.92—dc20
[B] 94-9480

Title page photo courtesy of Clive Barda.

1 3 5 7 9 8 6 4 2

Printed in the United States of America
on acid-free paper

For

R. C. F. B.

(1905-1983)

S. E. B.

(1919-1992)

Preface

This is a book of interviews. It attempts through a series of conversations to provide the reader with a mosaic of opinion on one of the more fascinating musical figures of our time, Leonard Bernstein. It does not attempt to cover in exhaustive detail every aspect of his career, but rather to give some inkling of how he worked, why he made the choices that he did, and why at the end of his life he felt that his single most cherished ambition—the creation of a serious masterwork—remained unrealized.

Organizing a book of interviews is not an easy task. First, responses to questions vary enormously—one interviewee will answer with two sentences; another will take two paragraphs. Second, there is the question of balancing one interview with another. Ultimately each interview has to stand on its own as a personal document that reflects as much on its source as on its subject.

To those interviewed—David Diamond, Christa Ludwig, Mstislav Rostropovich, Paul Myers, Harold Schonberg, John Mauceri, Justin Brown, Frederica Von Stade, Lukas Foss, Joan Peyser, Jonathan Miller, Jerry Hadley, Carol Lawrence, and members of the New York, Vienna, and Israel Philharmonics—I would like to extend my heartfelt thanks.

Bernstein was always a controversial figure, a man who throughout his early career with the New York Philharmonic was unable to get a good review in the New York press. At the same time, though, he elicited an unswerving loyalty from many of his colleagues, evidenced here by the contributions of some members of the orchestras with whom he worked. This book is, however, no catalogue of hagiographic tributes. It is rather an attempt to focus on the contradictions and the frustrations in

Bernstein, as well as being a celebration of his gifts. It is also an attempt to elucidate how in the last decade of his life he became to some extent the victim of his own celebrity.

As far as editing the interviews is concerned I have left speech patterns as close to the original as possible. I have, however, altered syntax in one or two cases where it was necessary for clarity.

I should like to thank for their valuable support and advice Sheldon Meyer and Leona Capeless, my editors in New York, as well as Bruce Phillips, with whom I originally discussed the idea for the book in Oxford. Also a note of thanks is due to Grant McLachlan, Michael Blake, and particularly David Chernaik, who helped with reading of the script. Most of all I would like to thank Dhun Manchershaw, who gave her consistent support during the project.

Contents

Introduction and Profile of a Musician

With the death of Leonard Bernstein on October 14, 1990, the musical world was deprived of one of its most vital forces. His death, which came a matter of days after he had announced his retirement from conducting, was received on both sides of the Atlantic with incredulity and shock. Bernstein had for so long been an integral part of the American musical scene that his prodigious talent and flamboyant presence tended to be taken for granted. When he died, America mourned its favorite musical son effusively. As one European critic noted: "When Karajan died the year before, corporations mourned. When Bernstein went, men and women wept in every Western metropolis as they recalled his extravagances and whistled tunes from *West Side Story*."

The multi-faceted Bernstein—conductor, composer, pianist, and proselytizer for classical music on television—was in many ways a pioneer among American musicians. The first American to conduct at La Scala—Cherubini's *Medea* in 1953 with Maria Callas—and the first to take over a major American orchestra when he became Music Director of the New York Philharmonic in 1958, Bernstein gave a new respectability to the idea of the American-born-and-trained musician. As Tim Page wrote in *Newsday* after Bernstein's death, his decision to make a career in America had "the same effect on our native musicians that Ralph Waldo Emerson's lecture 'The American Scholar' had on nineteenth-century literati. It was a declaration of independence. . . ."

With his astonishing versatility, Bernstein also became a giant of Broadway with such successes as *On the Town, Wonderful Town,* and the path-breaking *West Side Story*. His abundance of musical gifts was in fact one of his problems. Critics complained that Bernstein "spread himself too thinly" and characterized him as a musician who could never make up his mind whether his particular gifts were for Broadway or the concert hall, conducting or composing, or—something in which he excelled—as educator and purveyor of music to the masses. Colleagues were no less bewildered by the plethora of Bernsteins pulling in different directions, and Stravinsky referred to him as a musical "department store." Throughout his career the cognoscenti always wanted Bernstein to choose, to commit himself to one particular direction. Those who understood him best realized that this was impossible, that his very nature or identity was, to quote Ned Rorem, that of "Jack of All Trades."

Bernstein himself was always honest about the breadth of his desires. As Joan Peyser[1] pointed out in an interview for this book, he wanted everything—both sides of every coin: "He may have been homosexual, but he also wanted a family life; he wanted to be respected and respectable so he had a conducting career, but meanwhile he played on and off with Broadway. . . . " Peyser's biography was widely criticized, particularly in Britain, for having set out for public consumption some of the more intemperate, indulgent, and outrageous aspects of Bernstein's personality, be they on the podium or in his private life. What Peyser was doing was showing a human being who was not as "respectable" as his bourgeois public wanted or believed him to be.

Whether or not the man and the music are connected—Peyser believes that they are—there are certainly examples in Bernstein's music-making of wayward tempos, of attempting to imprint his own personality to such an extent that the music becomes distorted. Harold Schonberg, chief critic of the *New York Times* throughout much of Bernstein's tenure with the New York Philharmonic, sees Bernstein as a musician whose egocentricities interfered with his music-making: "He was just revelling in what he considered the music to be, and his tempos got very slow and they were extremely personal." Schonberg believes that Bernstein was a throwback, in a line that began with Wagner and continued through conductors like Bülow and Furtwängler. Bülow and Furtwängler may have used extreme fluctuation of tempo, but it is unlikely that they would have stretched a musical line to such an extent that the music threatened to disintegrate. A Bernstein interpretation, particularly in later years, would often risk doing this. When Bernstein conducted Elgar's *Enigma Variations* in London in the 1980s a member of the orchestra had the temerity (or the guts, depending on one's point of view) to question what the white-haired maestro was up to, going so far as to suggest that his ideas on *Enigma* were unmusical.

Bernstein's performances often involved the idea of not merely "interpreting" a piece of music but of "becoming" the composer during the period of performance, of, if you like, "recomposing" a particular work, something Bernstein spoke of at considerable length. In a description for the 1990 Leonard Bernstein Edition issued by Deutsche Grammophon, Bernstein elucidated as follows: "Perhaps the fact of being myself a composer, who works very hard (and in various styles), gives me the advantageous opportunity to identify more closely with the Mozarts, Beethovens, Mahlers and Stravinskys of this world, so that I can at certain points (usually of intense solitary study) feel that I have become whoever is my alter ego that day or week. At least I can occasionally reach one or the other on our private 'Hot Line'. . . ."

It was this sort of lofty pronouncement which exposed Bernstein so easily to ridicule and to caricature by his detractors. But there was another side to the conductor, a magnetic quality on the podium which Bernstein possessed perhaps to a greater degree than any of his contemporaries, and which produced the sort of full-blooded interpretations of composers like Mahler that were life-changing experiences for those privileged to hear them. Attending a Bernstein concert was always an event, with a considerable amount of glamor attached, and Bernstein looked the part of the grand old man of the podium *par excellence* in later years. Justin Brown, a Bernstein protégé, has said that there was a special quality attached to actually being *there*: "There was something he generated in the atmosphere of a concert that was completely memorable and that is not always recaptured on CDs." In fact not even the Unitel videos—Bernstein signed a lucrative contract with the Munich-based film company in the 1970s—manage to convey the particular crackle of a Bernstein performance.

In the early 1940s the Greek conductor, Dimitri Mitropoulos, who had in so many ways fired Bernstein's imagination as a young man, confided to the American composer David Diamond that he thought Bernstein a "genius boy," but was worried that he would burn himself out. Diamond remembers Bernstein from early on as a personality prone to excess, one example being his compulsive addiction to cigarettes: "I remember the cigarettes from as far back as Curtis, when Lenny was a student there. He always had a cigarette in his mouth and like everything with Lenny it was a competitive thing." Bernstein was, in fact, quite open about his excesses, whether related to smoking, drinking, or sexual promiscuity, once describing himself as "overcommitted on all fronts." Certainly a Bernstein who was celibate, vegetarian and teetotal would not have been Bernstein. It is possible that the way he lived his life was not so much self-destructive as a challenge directed at the frailty of the human condition. Norman Lebrecht, a London critic, has commented: "I don't think that the wilfulness of, for example, the smoking was in the depressive sense of trying to commit suicide. It was wilful in the far more elated sense of saying: 'Look at me; I can do this, I can get away with this. Can't I?'"

What is important, ultimately, is how and to what extent the

Bernstein lifestyle affected Bernstein the Musician. Particularly as he grew older, Bernstein seemed to find the discipline of composing increasingly irksome, and it is probable that his private life acted as a drain on his creative energies. Until the end of his life he spoke about the need to produce serious works that would last, that would give him some form of immortality. In November 1989, less than a year before he died, Bernstein told Yaacov Mishori, first horn of the Israel Philharmonic: " . . . I want to compose more. I don't feel happy that people will remember me because of *West Side Story*, even though I love the piece. I would rather people remembered me for my serious compositions." This supports the view that Bernstein was frustrated as a composer, that he was waiting to produce a masterpiece that never formed. Stephen Banfield, author of a recent book on Sondheim, has questioned whether Bernstein was ever able to "shut himself away for long periods, with discipline, to produce something big and careful." Mahler, Bernstein's great idol, did similar stints of conducting, but locked himself away in the summers in his Alpine hut and managed to produce his symphonies in a very disciplined way—one a year. It is possible, then, that had Bernstein lived his life slightly differently, he might have left a greater legacy of serious works. But speculation on this score is apt to be fruitless; one cannot say (as many critics did) that Bernstein *should* have made more of a commitment to serious composition, or, for that matter, to the piano, or the podium, or Broadway. One can only look at what he did achieve, in all these different fields, which was considerable. Joan Peyser has written in the *New Grove Dictionary of American Music*: "No musician of the 20th Century has ranged so wide. . . ."

Leonard Bernstein was born in 1918, in Lawrence, Massachusetts, to Russian-Jewish immigrant parents.[2] His father, Samuel Bernstein, who ran the Samuel Bernstein Hair Company, bought his son an expensive education—Boston Latin School followed by Harvard—and the young Bernstein in turn proved himself a brilliant if undisciplined student. Piano studies with Heinrich Gebhard—Boston's most celebrated teacher—and later Isabella Vengerova gave him the sort of technical equipment that, had he wanted, could have resulted in a career as a virtuoso pianist. He later made some superb recordings in this role, whether performing chamber music with the Juilliard Quartet or accompanying Christa Ludwig in Mahler *Lieder*.

Toward the end of the 1930s, Bernstein studied at the Curtis Institute under Fritz Reiner. Reiner, a strict disciplinarian, was considered with George Szell and Toscanini to be one of the great orchestral technicians of his time. His beat—the antithesis of the "choreographic" school of conducting—was minuscule. Reiner claimed that Nikisch had told him: "Don't wave your arms around; use your eyes to give cues." Bernstein, however, followed his own instincts on the podium, and audiences were treated to grand and exhilarating displays, with wild leaps and flailing arms, and the sort of eurhythmic movements that later prompted such critics as Virgil Thomson to write: "He shagged, he shimmied and, believe it or not, he bumped."

On November 14, 1943, Bernstein, who had been assistant conductor of the New York Philharmonic for two months, made his big breakthrough as a young conductor. Substituting for an indisposed Bruno Walter, he carried off what was probably the most sensational debut of any American conductor in this century. On page one of the *New York Times* there appeared the following:

> There are many variations of one of the six best stories in the world: the young corporal takes over the platoon when all the officers are down; the captain, with the dead admiral at his side, signals the fleet to go ahead; the young actress, fresh from Corinth or Ashtabula, steps into the star's role; the junior clerk, alone in the office, makes the instantaneous decision that saves

> the firm from ruin. The adventure of Leonard Bernstein, 25-year-old assistant conductor of the Philharmonic, who blithely mounted the podium at Carnegie Sunday afternoon when Conductor Bruno Walter became ill, belongs in the list . . . Mr. Bernstein had to have something approaching genius to make full use of his opportunity.

Bernstein's debut resulted not only in invitations from all over the country to conduct, but also in a two-week spot as a guest conductor with the New York Philharmonic, in the company of Igor Stravinsky, George Szell, and Pierre Monteux. Three months after his debut, Bernstein resigned his assistantship to Artur Rodzinski. It had been an unhappy period for the older conductor, who in his first year at the helm of the Philharmonic had been completely upstaged by his young assistant. Bernstein, however, had every reason for ebullience. The season of 1943–44 was in Bernstein's own estimate an "*annus mirabilis*" with regard to his career. Not only was he hailed as the most exciting young conductor of his generation; he also managed to compose a popular and adventurous ballet, *Fancy Free*, for Jerome Robbins, and his first musical, *On the Town*, with Betty Comden and Adolph Green. In a catalogue of "firsts," 1944 also saw the premiere of *Jeremiah*, his First Symphony.

For the first seventeen years of his career, Bernstein conducted without a stick. As a student of Koussevitzky at Tanglewood, Bernstein had asked his Russian mentor if he could dispense with the baton in favor of the use of his hands. He also managed to persuade Reiner—no mean task, considering Reiner's fearsome reputation among his students—to allow him to do away with a stick, by suggesting that the movements he made with his hands on the podium were connected to those made at the piano. This theory may have been based on his own particular experience at the time, but what is more likely is that Bernstein had made this choice because of watching Dimitri Mitropoulos conduct.

Mitropoulos, a Greek conductor of infallible memory and

dynamic physical presence, eschewed the baton and was probably the single most important influence on the young Bernstein. There were many points of identification, not least Mitropoulos's homosexuality, about which he was quite open, a courageous stance at that particular time in America. Mitropoulos was flamboyant, used his entire body on the podium, and reputedly enraged singers at the Metropolitan Opera with the unpredictability of his beat. His protégé Bernstein was also not noted for technical clarity.

Two years after his debut with the New York Philharmonic, Bernstein was offered—on his twenty-seventh birthday[3]—the musical directorship of the New York City Symphony. He succeeded Leopold Stokowski, a conductor whose performances were derided by many critics as exercises that were memorable for showmanship rather than musical substance. These were charges that would be levelled at Bernstein throughout much of his career.

Bernstein's tenure with the City Symphony, or the City Center Orchestra, as it was sometimes called, gave New York three remarkable seasons of twentieth-century works, and any number of world premieres, mixed with some of the more traditional repertoire. Harold Schonberg remembers these early concerts as attracting a young audience and being among the most stimulating New York had ever heard. He has written: "The atmosphere [at these performances] crackled like the rhythms in a Stravinsky ballet. Bernstein spoke the language of his young orchestra and his young audience." David Diamond also has fond memories of this particular period, whether of hearing the performance of his own Second Symphony (Bernstein championed a great many American composers at this time) or a work like Bartók's *Music for Strings, Percussion, and Celesta,* which Diamond said "got better and better" under Bernstein's enthusiastic direction. The young conductor gave his services during this period entirely without remuneration.

One of the reasons Bernstein may have committed himself to the New York City Symphony was that he sensed it to be a stepping stone to bigger and better things. In Boston the elderly Koussevitzky had told the board of trustees of the Boston Symphony Orchestra that he wished Bernstein to succeed him in the post of Music Director, one of America's most sought-after and prestigious positions. The trustees were not keen. First of all Bernstein was young—only twenty-nine—second, he was not an import from abroad, as were most major appointees by the five big American orchestras at that time, and third, he had a showbiz image that he was finding very difficult to live down, with *On the Town* currently showing to sold-out houses on Broadway. In addition, Bernstein, even at this stage in his career, was attracting some vitriolic reviews, and from some eminently respectable quarters. Virgil Thomson wrote, in 1946, in the *New York Herald Tribune*:

> With all the musical advantages he [Bernstein] has, he seems to have turned, in the last two years, even more firmly away from objective music making, and to have embraced a career of sheer vainglory. With every season his personal performance becomes more ostentatious, his musical one less convincing. There was a time when he used to forget occasionally and let the music speak. Nowadays he keeps it always like the towering Italian bandmasters of forty years ago, a vehicle for the waving about of hair, for the twisting of shoulders and torso, for the miming of facial expression of uncontrolled emotional states. If all this did not involve musical obfuscation, if it were merely the prima donna airs of a great artist, nobody would mind. But his conducting today, for all the skill and talent that lies behind it, reveals little except the consistent distortion of musical works, ancient and modern, into cartoons to illustrate the blithe career of a sort of musical Dick Tracy.

This sort of review can hardly have endeared Bernstein to an already skeptical Boston Symphony board, and the candidate eventually chosen by the board in 1949 was not Bernstein, but the affable Strasbourg-born Charles Münch, who specialized in the French repertoire, and whose conducting was considered the epitome of elegance and polish. Bernstein would have to wait another nine years before taking over one of the "Big Five" American orchestras.

2

From the beginning Bernstein had aspired to write music in the operatic or symphonic form. In 1944, his "*annus mirabilis,*" he made his first impression as a serious composer when he conducted the world premiere in Pittsburgh of his First Symphony, *Jeremiah.* Fritz Reiner, his conducting teacher from Curtis days, had invited him to perform the work. Reiner had not been overgenerous to Bernstein in the past, but possibly he now felt that Koussevitzky was gaining all the credit for the new *Wunderkind,* and that by inviting Bernstein he could share in some of the accolades.

Early performances of *Jeremiah,* scored for soprano and orchestra, were auspicious. Part of the work, the beautiful "Lamentation" section, which closes the piece, had been written as early as 1939, a time when there were none of the distractions (conducting or Broadway) which were to absorb Bernstein later. The work was finished in 1942, just less than a year before his celebrated conducting debut. From a critical point of view, *Jeremiah* was regarded as a work of promise by a young musician who was beginning to make his way, and one highly favorable review came from Paul Bowles in the *New York Herald Tribune.* Bowles wrote: "This work, which he calls *Jeremiah,* outranks every other symphonic product by any American composer of what is called the younger generation."

There can be no question of Bernstein's facility as a composer, or of his early talent. David Diamond, who met Bernstein when the latter was still a student, has commented as follows: "The question I have always asked myself was why, with Lenny so gifted he didn't end up with Nadia Boulanger. She would have turned him into the most famous man in the world, overnight." Had Bernstein gone to Boulanger he would have followed on the heels of some of America's brightest young men, including Aaron Copland, Walter Piston, Roy Harris, and Diamond himself. But Bernstein's attitude to study was never disciplined. At Harvard in the 1930s, where he sat in Walter Piston's composition and theory class, Piston had doubts about whether Bernstein would ever be a composer of serious import, because, as he explained, he was "always putting on a show or something." His extraordinary facility and his lack of discipline were qualities that would remain as characteristic of Bernstein at seventy as of Bernstein at eighteen.

In 1949 Bernstein completed his Second Symphony, *The Age of Anxiety*, based on Auden's poem of the same name. The music had been written during a time of hectic touring and concert appearances, and by the composer's own admission had been created partly in the ambience of international airports. If *Jeremiah* had been greeted warmly as a promising first endeavor in the symphonic form, *The Age of Anxiety* was regarded as a disappointment. Highly eclectic in content, the work was a somewhat naive attempt to infuse a classical form with commercial art. As in so many of his later works—*Mass*, for example—Bernstein was trying here to reconcile serious and popular culture. This inevitably led to accusations that he was trivializing his talent, and prompted *New York Times* critic Olin Downes to write that *The Age of Anxiety* was "wholly exterior in style, ingeniously constructed, effectively orchestrated, and a triumph of superficiality."

If Bernstein's serious works were coming in for a hammering from the critics at this time, his work on Broadway, beginning with the earlier productions of *On the Town* and *Wonderful Town*, both of which had been popular shows, culminated in 1957 in his greatest success as a composer for the musical theatre, with *West Side Story*. The only hiccup in his string of Broadway triumphs was the ill-fated first production of *Candide*, which opened the year before.

As early as 1950 playwright Lillian Hellman had suggested to Bernstein a collaboration on Voltaire's masterpiece, and several years and several lyricists later, on December 1, 1956, *Candide* had its New York premiere. Bernstein's witty, irreverent score, his most European to date, ran in its original production for fewer than eighty performances on Broadway and *Candide* became the most rewritten work (in terms of the book) in the history of the musical. There were so many different collaborators that as Jonathan Miller, who much later directed the work at Scottish Opera, has commented: "Practically everyone in America seemed to have done something. . . ." After 1956 the work had a checkered career, and although it received a number of revivals, most of them more successful than the original, it found its final form not in a staged production, but in Bernstein's own concert recording for Deutsche Grammophon in 1989. John Mauceri, the American conductor, for many years one of Bernstein's assistants, reorganized much of the piece (as with a later work, the opera *A Quiet Place*) and incorporated some memorable songs which had been cut from the original.

Following *Candide*—as always Bernstein had been working frenetically on several different projects at once—he produced *West Side Story* (1957). Conceived, directed, and choreographed by Jerome Robbins, with a book by Arthur Laurents and lyrics by Stephen Sondheim, *West Side Story* was one of the most successful collaborations Broadway ever achieved. Ironically, the work had difficulty finding a producer, and Cheryl Crawford, who had initially taken on the show, pulled out. Carol Lawrence, who played the role of Maria in the original production, has commented: "She [Cheryl Crawford] thought it would be critically acclaimed but that the American public would never buy it. . . ." Hal Prince, who had also initially turned the work down, eventually undertook the production with Bobby Griffith, and the rest is musical theatre history.

Rehearsals for *West Side Story* took place under ruthlessly tight control by director/choreographer Jerome Robbins. Robbins applied Stanislavsky's "Method" wherever and whenever possible, and reportedly had the cast in tears for much of the time. Members of the two opposing gangs in *West Side Story*, the Jets and the Sharks, were not allowed to communicate with each other once they arrived at rehearsal, and everybody, including Bernstein and the other collaborators, deferred to Robbins. His techniques paid off, and the new work succeeded spectacularly in its integration of drama, music, and dance, something its predecessors *On the Town* and *Wonderful Town* had never quite managed. If Bernstein had over the years resented his lack of recognition as a serious composer, his gifts as a writer for the musical theatre were fully vindicated by *West Side Story*. Some critics preferred his jazzy, kinetic rhythms, used to underscore Robbins's choreography, to his lyrical writing for Tony and Maria, the lovers, but these were small quibbles. Most people felt that Bernstein with *West Side Story* had produced his most memorable and also his most universal score. Later, and particularly after the composer's controversial 1985 opera-star recording for Deutsche Grammophon, there would be argument as to whether *West Side Story* was most at home on Broadway or in the opera house. Stephen Banfield, an authority on the work of Sondheim, has commented: "With Bernstein I think the lines between what constitutes a musical and what constitutes opera were not always clearly defined."

Bernstein himself, however, in spite of his later recording of the work, did not make operatic claims for *West Side Story*. He commented that at the end of the show, when Tony has been shot and Maria picks up the gun, "the music stops and we talk it." Bernstein tried at this point—unsuccessfully—to write an aria, but instead the most powerful moment in the show—and the ultimate "message" of *West Side Story*—remains Maria's final speech.

3

In 1954, three years before *West Side Story* opened, Bernstein embarked on an entirely new stage of his career, and one in which he would surpass himself—the *Omnibus* television broadcasts. *Omnibus* was a show financed by the Ford Foundation, and it brought Bernstein's talents as an educator and proselytizer for classical music into the living rooms of millions of Americans. Joan Peyser has written as follows on the *Omnibus* series: " . . . *Omnibus* shows dealt with what makes jazz jazz, why an orchestra needs a conductor, what makes Bach Bach, why modern music sounds so strange, what this distinctive form called musical comedy is, and what makes opera grand." Where Bernstein scored was in bringing classical music—previously thought inaccessible—to the widest possible American public. His television programs managed to interest the man in the street as well as the academy, and his audiences for *Omnibus* and later *The Young People's Concerts* were treated to a wide range of music, with a commentary that was sophisticated and never patronizing.

By 1958, Bernstein found himself to be in an enviable position. First, *West Side Story* had opened the previous year and Bernstein had written in a letter to David Diamond that the show was "provisionally a smash-hit." Second, its composer was about to secure the music directorship of one of America's greatest orchestras. It was a double-act that few could follow. As

Paul Myers, for some twenty years a producer for CBS records, has pointed out: "It was a difficult hurdle for him; he had been the great Broadway idol, who had made the transition from Broadway to the Concert Platform . . . who else has done it?"

When Bernstein inherited the New York Philharmonic from Dimitri Mitropoulos it was not under the happiest of circumstances. Mitropoulos and Bernstein had been co-directors of the orchestra, in 1957, after which the older man was pushed out and the younger conductor given the post the following year. The management of the NYPO predicted, correctly, that the youthful glamour of a Bernstein would cause audience and subscription levels, currently at a low ebb, to rise. In addition, the Philharmonic that Bernstein inherited from Mitropoulos was an orchestra that was badly disciplined and suffering from poor morale. It had been the subject of a blast in the *New York Times* from Howard Taubman in a now famous essay entitled: "The Philharmonic—What's Wrong with It and Why."

Among problems that Taubman enumerated were haphazard programming, an unimaginative choice of guest conductors, and Mitropoulos's apparent inability to control the orchestra. It should be pointed out that disciplinary problems were not confined to the Greek conductor's tenure; the Philharmonic had a reputation for demolishing guest conductors, and great musicians who had suffered at the orchestra's hands in the past included Otto Klemperer and Bruno Walter. Taubman's essay had a tremendous impact on Bernstein's career. As he later admitted, his article "gave the Philharmonic courage to give the post to an American—and a young man."

Bernstein brought an excitment and a glamour to the organization. He also did much during his eleven-year tenure for the morale and the financial status of the orchestra, with increased revenue from recordings and television, tours to South America, the Soviet Union, and Europe, and a longer season which enabled the players to become full-time employees, under contract. In spite of all of this and the esteem and affection in which he was held by most of the orchestra members—he was given the title "Laureate Conductor" when he left the Philharmonic in 1969—there were doubts in the minds of some critics as to the quality of what he was producing.

If Howard Taubman's article in the *New York Times* had helped Bernstein into the top job at the New York Philharmonic and provided him with largely favorable reviews in his first two seasons (1958–60), the man who succeeded Taubman was to prove Bernstein's severest detractor. Harold Schonberg wrote as follows after one of the concerts in the 1960–61 season:

> Mr. Bernstein did a pretty good job upstaging Mr. [Sviatoslav] Richter . . . He made almost as much noise on the podium as his colleague did on the piano. Such foot stompings have not been heard since the Fifty-fifth Division was on parade . . . Obviously Mr. Bernstein was exhilarated. In the "Battle of the Huns" he did everything but ride a horse to battle. Towards the end of the Liszt Concerto, he rose vertically in the air, à la Nijinsky, and hovered there a good fifteen seconds by the clock. His footwork was magnificent last night. But one did wish that there had been more music and less exhilaration.

Schonberg's reviews may have been harsh, but there was some truth in them. Bernstein did upstage soloists, and his mannerisms on the podium often tended to obscure the music. His more personal interpretations, and particularly his liberal use of rubato, suggested a conductor in the nineteenth-century tradition rather than a representative of the more literal modern school.

In 1961 the New York Philharmonic board, ignoring Schonberg's admonitions, extended Bernstein's contract for another seven years. Surprisingly, only twelve weeks of concerts a year were stipulated, an arrangement which must have suited Bernstein, as it gave him time for composition. He had another reason for ebullience: audience figures had been turned around since the Mitropoulos era, with his concerts playing to packed houses, first at Carnegie Hall, and then at the new Philharmonic (later Avery Fisher) Hall, which opened as part of Lincoln Center in 1962.

Throughout his tenure at the Philharmonic in the 1960s, Bernstein programmed a good deal of contemporary American music, including some of the dodecaphonic specialists, but he was never entirely comfortable performing music of the avant-garde, mainly because of his own beliefs in tonality. During his sabbatical from the Philharmonic, in the 1964–65 season, he experimented a good deal with the twelve-tone system in his own composition—such eminent figures as Copland and Stravinsky were doing likewise—only to come up with the *Chichester Psalms,* a work firmly rooted in tonality, which he described as "simple and tonal and tuneful and pure B flat as any piece you can think of."

In 1966, Bernstein held a press conference to announce his intention to leave the New York Philharmonic as Music Director after the 1968–69 season. According to one writer it looked as though "a chief of state was about to step down before his time." Bernstein's reason, given in the form of a written statement to a room packed with reporters, was his compulsion to compose. The only two major works that Bernstein had come up with during his entire tenure at the Philharmonic were the *Chichester Psalms,* written during his sabbatical, and *Kaddish,* the Third Symphony, completed a year before.

On August 25, 1968, Bernstein would be fifty, the age of Gustav Mahler when he died. Much was made of the similarities between Bernstein and Mahler—mainly by Bernstein himself—particularly of the complications of being both composer and conductor. Mahler was neurotic, arrogant, insecure, tortured over questions of Judaism versus Catholicism (to which he converted, for career purposes), and frustrated by the fact that his conducting career left him a minimum of time for what he regarded as his main mission in life, his creative work. All of these attributes could, to a greater or lesser extent, be applied to Bernstein. However, there was one major and all-important difference between the two men, which Bernstein realized only too well: by the time of Mahler's premature death, aged fifty, he had left behind a legacy of symphonic masterpieces as well as some of the most beautiful song cycles of the twentieth century, *Das Lied von der Erde, Des Knaben Wunderhorn,* and the *Kindertotenlieder* among them. Bernstein, rather than acknowledging his own creative achievements in the somewhat different genre of the musical theatre field, was obsessed with the idea of composing a serious symphonic or operatic masterpiece, the only satisfactory way he saw of gaining some form of immortality.

4

In 1971, in spite of the battering that his serious compositions were receiving in the press, Bernstein—now free from his conducting duties at the New York Philharmonic—produced what was to be his most eclectic piece yet. *Mass,* written for the opening of the Kennedy Center, was Bernstein's attempt to commemorate what he perceived as the "Camelot" of John Kennedy's presidency. Subtitled "A Theater Piece for Singers, Players and Dancers," *Mass* involves a large performing force—some two hundred—and revolves around the central figure of a celebrant, who, like the narrator in *Kaddish,* is a self-portrait of the composer. If *Kaddish* had suffered from charges of eclecticism, *Mass* would go a stage further. Bernstein himself described the work as containing "the whole Latin Mass, symphonic music, plus pop-sounds and blues." The piece was savaged by the critics. Harold Schonberg wrote that *Mass* was "the greatest melange of styles since the ladies' magazine recipe for steak fried in peanut butter and marshmallow sauce." John Simon, in *New York Magazine,* commented: "The trouble is not so much that it is eclectic, as that it is banal, inappropriate and rather vulgar."

Following *Mass,* Bernstein undertook a project that would once more satisfy his compulsion as an educator, his "quasi-rabbinical instinct to teach," as well as being an academic honor of the highest order—the Harvard Eliot Norton Lectures. Begun in 1973, Bernstein used the lectures—an attempt to apply linguistic concepts to musical analysis—as a platform for the justification of his own beliefs in tonality. The Bernstein *credo* enraged many composers who had committed themselves to the dodecaphonic school, and Bernstein came in for some angry attacks from musicians of the avant-garde. Another criticism was that Bernstein's approach, with all its attendant showbiz paraphernalia, was not as esoteric as what had gone before. Previous occupants of the Norton chair had included T. S. Eliot, e.e. cummings, and Igor Stravinsky, and none of Bernstein's predecessors had turned the lectures into the sort of multimedia event that characterized his appearances.

The crucial factor in Bernstein's decision to leave the New York Philharmonic, in 1969, may have been his urge to compose, but there were other reasons. In 1966, Bernstein began what was to become an extraordinary musical partnership, with the Vienna Philharmonic, conducting *Falstaff* at the Vienna *Staatsoper.* This marked a turning-point in his career. *Falstaff* received rave reviews from the Viennese press and set the tone for a critical change of heart on both sides of the Atlantic. Rudolph Klein wrote from Vienna for the *New York Times:*

> Since the departure of Herbert von Karajan from the Vienna State Opera, no conductor has been so extolled in this house as was Leonard Bernstein for the premiere of his production of the opera *Falstaff.* . . . He certainly deserved the ovations: his

> work achieved the maximum both on stage and in the orchestra. Moreover one never had the impression that here was a dictator issuing commands with an iron will. Quite the contrary, each musical phrase came forth as improvised, as if of itself, without any compulsion. . . .

Bernstein's success in Vienna, and, more particularly, with some of the Viennese critics, contrasted sharply with the rough ride he had had in New York. Vienna, Mahler's city, must have appeared a seductive proposition, and toward the end of the 1960s Bernstein began attenuating his ties to America. Exactly why the Vienna Philharmonic took to him in the way it did is a fascinating question. The Viennese musicians appeared to have a love affair with Bernstein. One member of the orchestra commented: "Bernstein opened all doors with us because he had the courage to translate all his feelings into movements without restraint." More cynical commentators suggested that it was a shrewd commercial move on the part of the orchestra to cultivate a figure with Bernstein's popular appeal. Whatever the case, Bernstein's success in Vienna was phenomenal and he was accorded such honors as leading Beethoven's *Missa Solemnis* for the hundredth anniversary of the opera house. In the spring of 1968, he performed and recorded *Der Rosenkavalier* in Vienna.

It is one of the anomalies of Bernstein's career that as a Jew he should have enjoyed the adulation of what is reputed to be one of the most anti-Semitic cities in Europe. Bernstein made Vienna virtually his powerbase throughout the 1970s and 1980s, signing a contract with the Munich-based film company Unitel and establishing himself as a leading artist on the roster of Deutsche Grammophon. His reasons for all of this were probably manifold. In one sense he perhaps saw himself as a force for good, and to some extent he was, taking the Vienna Philharmonic on its first tour of Israel, where trees were planted to the memory of Jewish martyrs, and restoring the bust of Mahler—removed by Hitler—to its place at the Vienna *Staatsoper*. All the signs, however, pointed to the fact that anti-Semitism in Austria was still alive and well, and this was confirmed in the 1980s by the election of Kurt Waldheim to the office of President, the Austrian electorate remaining undeterred by Waldheim's Nazi past. When Bernstein, a man who always liked to appear socially conscious, was asked whether he would be returning to conduct in Vienna, he replied that the musicians were his "Brüderlein" and that he could not abandon them.

5

Throughout the 1960s Bernstein had been a prolific recording artist for CBS. By the early 1970s his recording career was on something of a plateau. It was at this time that he pushed for a release from his American contract and a move to Deutsche Grammophon. The German company believed that signing Bernstein would give them a foothold in the American market, and their calculations paid off handsomely. When in 1985 Bernstein recorded *West Side Story* for the famous yellow label, he provided Deutsche Grammophon with their best-selling record ever. Through his lucrative contract with Unitel, Bernstein also arranged to have most of his work filmed. This meant that every time he mounted the podium he was paid for the concert, the recording, and the video. He chose now to have his repertoire recorded "live," in order to capture as much of the excitement and electricity of a concert performance as possible.

By the beginning of the 1980s Bernstein, in spite of all the time he had taken off from conducting—he took the whole of 1980 off in order to compose—was wondering whether the serious masterpieces he had always hoped to write were ever going to materialize. In 1980 he produced his *Divertimento,* written for the Boston Symphony Orchestra's hundredth anniversary, a frothy, insubstantial piece. The *New York Post* remarked:

"The whole seemed a trifle for such an occasion, unmemorable even as light music."

The preceding years had not been an easy period for Bernstein. Felicia, his wife, had died in 1978, after a battle with cancer. Prior to her illness, Bernstein had begun a period of separation from her—he left her to live temporarily with a male lover—and following her death he reportedly became overwhelmed with guilt. Nevertheless, Bernstein the conductor and composer kept going at a frenetic pace. In 1980 he met the young writer Stephen Wadsworth, who produced the libretto for the work that Bernstein had been promising the American public—and himself—for so long, the opera *A Quiet Place.* Bernstein said of the piece: "It is not yet another attempt to take the American musical a step further. It is qualitively different. It is opera written for the opera house." Besides Wadsworth, there was one other figure who played a pivotal role in the work—John Mauceri. Mauceri, after the first performances in Houston, re-organized large sections of the opera, with the composer's approval.

Bernstein finished *A Quiet Place* in 1983. He put enormous energy into the project, often working with his young librettist around the clock. Musically, the work was a mixture of everything from jazz to twelve tone and incorporated *Trouble in Tahiti,* an earlier one-act opera from the 1950s. Bernstein's new opera opened on June 17, 1983, at Jones Hall in Houston. The response, both critical and public, was lukewarm. What was particularly hurtful was the reaction by some critics to the nature of the characters in *A Quiet Place,* who were drawn from Bernstein's own life. Bernard Holland, in the *New York Times,* asked "why this immensely talented man has allowed himself to squander his fury and intensity on characters like Sam, his grown children, and the children's bisexual lover. Their sufferings repel rather than move."

Although the opera eventually had a respectable hearing in

Europe, its initial reception was a bitter disappointment to Bernstein. John Mauceri, who conducted the La Scala premiere of the work, has commented: "The response was negative, and swift and clinical, lethally so. When Lenny was writing *A Quiet Place* he was so happy he thought he would only write operas for the rest of his life. And of course he never wrote another one." Bernstein's last years as a composer were occupied by projects of a much smaller scale, like *Arias and Barcarolles,* a cycle of songs (with texts mainly by the composer), which had its premiere in May 1988.

Toward the end of the 1980s Bernstein was still one of the most sought-after conductors, but he was beginning to find his podium activities increasingly strenuous. He had suffered for several years from progressive emphysema, a disease of the lungs, brought on by his life-long addiction to cigarettes. Smoking was one of several areas of his life that he seemed unable to control. The end of 1989 found him in London, seriously ill,[4] yet still infusing life into the score of one of his own more problematic offspring, *Candide.* On Christmas Day of the same year, in spite of his ailing health, he directed a concert in Berlin to mark the crumbling of the Berlin Wall. In a political gesture characteristic of Bernstein, he changed the words of Beethoven's *Ode to Joy* from "Freude" to "Freiheit."

At the end of his life, Bernstein seemed to enjoy his work with young people more than any other single activity. In 1986 he commented: "I don't feel vindicated as a composer, but I do as a teacher." He perhaps left his greatest legacy to those students and protégés who were lucky enough to work with him at places like Tanglewood and Schleswig-Holstein. One protégé, Justin Brown, remembers that Bernstein's main role was as an inspirational musical force: "He saw his function at Tanglewood, I think quite rightly, as being to talk about music. There were other people telling you how many beats to beat in the bar. . . ." In the last year of his life, 1990, Bernstein set up the

Sapporo Festival in Japan, an attempt to establish a "Tanglewood of the East." Only weeks later, on August 19, 1990, he conducted his last concert, appropriately enough the Serge and Olga Koussevitzky Memorial Concert, at Tanglewood. Included on the program were Four Interludes from Britten's opera *Peter Grimes* and Beethoven's Seventh Symphony.

Tim Page, a long-time Bernstein watcher, wrote:

> He nearly broke down in the middle of the Beethoven and conducted much of the third movement leaning against the back of the podium, gasping for breath. Yet he kept on, clearly determined to make this the performance of a lifetime, bringing out an extraordinary amount of detail from the score while never losing track of its central pulse. Rarely have the symphony's silences seemed so portentous, its climaxes so exultant . . . .

Bernstein cancelled all conducting engagements following this concert, announcing his retirement from conducting a matter of days before he died. But he was still making plans for future projects, and the hope was still alive that he would continue to compose. In a rather poignant interview (given while recording *Candide* in London in 1989) he told Edward Seckerson of *Gramophone* magazine:

> You know, the feeling gets stronger and stronger that in whatever few years I am granted to remain on this earth, I really should spend most of it composing. Because a great many people can do Bruckner's Twelfth (*sic*) well, and some can do it magnificently, and I'm not really needed on this earth for another *Ring* cycle, or another *Magic Flute*, or another whatever. Really. But nobody, for better or worse, can write my music except me.

NOTES

1. Joan Peyser is the author of a biography of Bernstein, published in 1987.

2. Bernstein was originally named Louis Bernstein at the insistence of his maternal grandmother. Later at the age of sixteen he legally changed his name to Leonard.

3. This may have served to confirm Bernstein's vague beliefs in some sort of numerology, because his appointment as assistant-conductor of the New York Philharmonic had also come on his birthday, August 25, his twenty-fifth.

4. Bernstein had a cancerous growth on the pleural sac of the lungs.

Conversations About Bernstein

The Conductor:

"He is a specialist in the clenched fist, the hip swivel, the pelvic thrust, the levitation effect that makes him hover in the air in defiance of the laws of gravity, the uppercut, the haymaker."

Harold Schonberg in *The Great Conductors.*

The Composer:

"He has left music none of which is dull, much of which is mediocre and some of which is imaginative, skilful and beautiful. There is rightly much controversy as to its lasting value."

Leonard Bernstein (on George Gershwin).

The Educator:

"I grew up on Bernstein. When I was a kid in Indiana I watched the telecasts of *The Young People's Concerts and Omnibus,* and I think a whole generation of music lovers and professionals, including myself, are there, at least in part, because of this man."

Jon Deak, Associate-Principal Double Bass,
New York Philharmonic.

I
Composers on Bernstein

Lukas Foss

Lukas Foss, the American composer, was born in Berlin on August 15, 1922. He graduated from the Curtis Institute in Philadelphia in 1940, having studied conducting with Fritz Reiner, piano with Isabella Vengerova, and composition with Rosario Scalero. It was at Curtis that he first met Bernstein, who also attended Reiner's conducting class. Later both men would study with another great conductor, Serge Koussevitzky, at Tanglewood. Foss's compositions show influences of Hindemith (with whom he studied), Copland, and Stravinsky, but these are all assimilated into his own definitive and individual style. In the 1960s he emerged as one of the more important influences on American experimental music. Like Bernstein, he has pursued successfully a career as composer, conductor, and pianist. Many of Foss's works were performed by Bernstein, including *Song of Songs* (1946), *Introductions and Good-byes* (1959), *Time Cycle* (1960), and *Baroque Variations* (1967).

Could we begin with your attendance at Fritz Reiner's conducting class, of which Bernstein was also a member, in 1939? Was Reiner as strict and demanding as he purported to be?

Well, Reiner was a stern teacher, and he could certainly be tough; he was not someone you could be very familiar with. I remember in his class once when Lenny called him "Fritz," Reiner shot back: "Yes, Mr. Bernstein?" But I think Lenny admired him. The first time I really got to know Lenny well was when the Curtis Institute sent us by train to Chicago, where Reiner was conducting *Rosenkavalier*. We were sent to witness the rehearsals and the opera as a form of extra study. That was when Lenny and I really became friends. He introduced me to the music of Gershwin and to jazz, which until then had not been important in my life, and helped me to understand what they were about. Of course we also had the same piano teacher, [Isabella] Vengerova, and then later we both studied with Koussevitzky.

Was Bernstein happy at Curtis? He has been quoted as saying that as a Harvard graduate he felt uncomfortable being the only "university type" at the school.

To me Lenny was always able to cope with *everything*, I don't recall his being uncomfortable at all. He was very sophisticated already, whereas I was like a little boy. I was only fifteen years old, very immature, and Lenny was almost twenty and like an older brother to me. He seemed so at ease that he was not even like a student; it was almost as if he was in charge!

Bernstein even at this stage had apparently developed extraordinary sight-reading abilities. Do you remember instances of this in Reiner's class?

Oh yes, Lenny's sight-reading was truly amazing, he could sight-read anything. And he also could *memorize* anything. It was an aural memory. It was not like Mitropoulos, who had a very visual memory.

Would you say that out of all of Bernstein's influences as a conductor Mitropoulos was the most important, in spite of the fact that Bernstein never studied with him?

It is interesting you should say that; I would have guessed that Koussevitzky was the most important. You know part of what Lenny got from Koussevitzky was this feeling of life in a performance; every performance was like a *first* performance. That's one of the ways in which Reiner and Koussevitzky differed.

With Reiner sometimes you had the feeling that he could do a concert in his sleep, whereas with Koussevitzky you had the feeling that he was doing, for example, the Tchaikowsky Fourth Symphony for the *first* time, even though it was actually the hundredth time. And that you felt with Lenny too. He put absolutely everything he had into a performance. That was why watching Lenny at Tanglewood was very much like watching Koussevitzky (from this particular point of view). Of course Koussevitzky always acted like a father to his students. He was totally different in attitude from Reiner in this way, as a teacher. Reiner would be stern whereas Koussevitzky was always very loving.

Bernstein's relationship with his own father appears to have been very difficult. (Bernstein's father had no idea of the nature of his son's talent, and envisaged him spending his days playing the piano in a bar or cocktail lounge.) Do you remember this?

Well, I remember a concert at Town Hall in New York [February 1943] where Lenny played the Copland *Variations.*[1] After huge applause had broken out in the hall there was this man—it was Lenny's father—leaning back into my ear and saying: "That's all very good, but where's the money?" You see, Lenny's father wanted him to continue in his beauty parlor business [Sam Bernstein ran the Samuel Bernstein Hair Company], and to continue with *his* work. So Lenny had to persuade him about his music. I think also Lenny wanted to prove to his father that with his own career he could also be good in business, which of course he did.

In the early 1940s Koussevitzky suggested to Bernstein that in the interests of his career he ought to think about changing his name. Do you remember this?

Yes. And I'm very glad that Lenny resisted. I think that at the time Bernstein was considered a lowbrow Jewish name and Koussevitzky was aware of the importance of the conductor's image. [Koussevitzky, who suggested to Bernstein the name

"Leonard S. Burns," had himself converted from Judaism while still in Russia in order to further his career.] I was told that Ormandy also changed his name, he changed it from Eugen Blau to Eugene Ormandy! I think conductors at that time were very name-conscious. Some of them wanted to try to find a name that would sound like Toscanini. It is the same thing with actors—actors also consider what name will be appropriate for their image.

Moving on from the days of Curtis and Tanglewood, you and Bernstein performed together on numerous occasions over the years. As a conductor, how easy did you find it to follow him?

Well, generally he was not difficult to follow. But I think at times when he was very free, or in very slow tempi, when he would not subdivide, then he *could* be difficult to follow. As he became older the tempi tended to become slower. I remember one instance of this particularly well. He came to hear a concert of mine where I conducted his *Chichester Psalms*—it was in Milwaukee [in 1985] and I was doing a Bernstein Festival there. Lenny came backstage and said: "I love the way you did *Dybbuk* [a ballet choreographed by Robbins which Bernstein later turned into an orchestral suite] but frankly, what you did to the *Chichester Psalms*, the tempi were all *wrong*." I said: "But Lenny, my tempi are exactly yours; I was so careful about that. . . ."

"Oh *no*, you must listen to my record."

"Well that's exactly what I did; I listened to your record, I took *exactly* the tempi of your record."

Then suddenly he looks at me: "Oh! I understand. You listened to my *first* record!" You see, his recent record was infinitely slower than his first one, so that shows what happened to his conception of tempi in some cases.

What do you think prompted this slowing down?

In a sense that came from Lenny's desire to really pump the most out of the music, to milk it, to get everything out of it that was in it. Sometimes he would do that by driving home the point, by being totally emphatic about every detail. I think that is how the tempi became slower, I don't think it had to do with failing health or anything like that. If you want to make sure that people hear the detail in a piece then you slow things down.

Do you think that there were educational impulses at work with Bernstein's music-making?

Yes. He wanted to really *communicate* what was in the music. I remember when he did Elgar's *Enigma Variations* in England, he decided to show the English that it was "*much* greater than they thought!" So he did it slower to drive home the point!

During his time with the New York Philharmonic (and also when he directed the New York City Center Orchestra) he programed much new music, particularly American composers, including yourself. How did he approach the first performance of a new work?

Well, he would put, how can I say, a loving searchlight on the music. I remember when he did my *Time Cycle*, no, actually it was the *Baroque Variations*, he called me up at eleven at night and said: "Lukas, I don't understand your music anymore. Can you come over here and explain it to me?" I said: "What? Now, at 11 p.m.?" He said: "Why not?"

So I went over to Lenny's place, took my score, and we worked together until two in the morning. First I explained to him what I had in mind, and then by the time it was 1:30 or so *he* explained to *me* what I had in mind! Now Lenny had a rehearsal of the work the following morning, and he came with some thirty pages of notes which he had written out about how he was going to rehearse it. He must not have gone to bed. He had actually taken the trouble to write down how he was going to prepare the piece.

In the rehearsal he was totally on top of it. *Baroque Variations* is a very problematic work. There is something in it that is very tricky; the players don't stop playing, they become inaudible, and then they go back into audibility and then back again and so on. Lenny knew just how to rehearse it, and he got amazing results from the players. I remember he turned to them at one point and said: "Did you hear what you just did? Do you realize you're the first people who have ever made this sound?" Well, and the New York Philharmonic took wings! What Lenny invested into other people's music, his colleagues' music, was extraordinary; there was a complete unselfishness in the way he did this.

To move on to Bernstein's own compositions, he has often been accused of being too eclectic. Would you agree?

Well, first of all, who isn't eclectic? Everybody is eclectic. What we have to do is to look at the influences and what he did with them. There was for instance a Hindemith influence, even though he never studied with Hindemith. In the Clarinet Sonata [composed in 1942] you will find influences of Hindemith and Copland. Eventually the Copland influence triumphed, because Lenny became more interested in American music and he got away from playing with this kind of neo-classicism. Then he also discovered Stravinsky. And then there is also Gershwin of course, the jazz and Broadway side, which left its mark. The important thing about musical influences is that when you use them in a work you then have to make that work your own. And Lenny did that. Throughout his compositions he did that; even the early works were *Bernstein.*

He often seemed to use material from other composers with a minimum of alteration. There are instances of this in for example the Chichester Psalms, *where he quotes from Beethoven (a theme from the* Pastoral Symphony*).*

Yes, Lenny loved to quote, and his quotes were unashamed and wonderful. He also quotes *me* in the *Chichester Psalms.* He took from my *Psalms* something which became very prominent in his work, and he even told me the history of it; it was quite amusing. He said to me: "Well, Lukas, here's what happened. I liked that tune in your *Psalms,* and I used it in *The Skin of Our Teeth* [a musical version of Thornton Wilder's play, which Bernstein had worked on during his sabbatical from the New York Philharmonic in 1964–65, but which never saw the light of day]. And when that didn't work out, I used the material from *The Skin of Our Teeth* for my *Chichester Psalms* so it was like the criminal re-visiting the scene of the crime! Your tune found its way back to the same words!" [In fact, Foss's words were in English whereas Bernstein's were in Hebrew.]

You wrote, in the preface to A Complete Catalogue of Bernstein's Works *by Jack Gottlieb, that Bernstein's music has the quality of "instant communication." Why do you think this is?*
Well, with Lenny, writing music was never an intellectual exercise. For example, he never wrote in the way that the Schoenberg-Webern school did, even though he may have written a very small amount of twelve-tone stuff. Lenny never wrote "paper" music; it was always for *immediate* communication. That's what he was about. As far as the atonal parts in his works are concerned they are just moments really, and tonality was absolutely important and essential to him, as was melody. He always said: "I believe in melody." Don't forget, a lot of music of that time had no melody whatsoever. But Lenny was very open-minded. For example, when I became wildly avant-garde, he did not turn away from my music. On the contrary, he would ask me to explain it to him.

You don't think that as a conductor he favored tonally based composers?
Lenny favored the music he loved. But he deserves special credit for not overlooking or rejecting music he did not take to. He was interested in all music.

After West Side Story *had its first highly successful run on Broadway (1957), do you think Bernstein became fixed in the public consciousness as a*

"Broadway Composer"? Do you think that hindered recognition of so-called serious works?

I remember Lenny said to me once: "I guess I'm not a serious composer." At least he meant that that was what people thought, that he wasn't a serious composer. What would happen was that sometimes people would *not* take him so seriously, and then one would get the wrong image, that image would be made into a cliché and so on. You know, in our age people always want to simplify things, to pigeon-hole artists, and I think that is wrong. For instance, in literature two of the most serious writers, Kafka and Beckett, are also the greatest humorists. And with Lenny I must tell you that I think his most serious work, also one of my favorite works, is *West Side Story*! And I will tell you why, because it is *passionate*. Of course, I am in the minority there; most people consider *West Side Story* to be "Broadway," but I think it is extremely serious.

Many of his compositions have some kind of literary connection or starting point, for example, the Serenade *after Plato, or* The Age of Anxiety *based on Auden's poem. How important do you think a literary stimulus was to his composition?*

Well, of course, Lenny was a *great* reader. And anybody who is well versed in literature will find inspiration in literary works. The various arts will always influence each other. Probably the reason he had so much success with his collaborations in the music theatre was that he was fired by the intrusion of the other arts, that they stimulated his imagination. I would say Lenny was the most well-read composer I have ever met; for example, he knew a large amount of poetry from memory. And conductors I think are generally not on that level at all. They are not even on the level of composers in this way; they haven't got that kind of all-round curiosity in them. Conductors are more occupied with their own image!

Looking for a moment at Bernstein's three symphonies, Jeremiah *seemed to promise much that* The Age of Anxiety *and* Kaddish *failed to deliver. Would you agree?*

I wouldn't say that about *Age of Anxiety,* but possibly *Kaddish* didn't make it in the same way as the other two works. But you know, some works catch on and some don't and that's true of any composer. And it doesn't necessarily have anything to do with the quality of the work! *Age of Anxiety* I know particularly well from having played it [the piano part] almost my whole life. I recorded it with Lenny, first with the New York Philharmonic [for Columbia Records in 1950, a year after the work was finished], and then we recorded it again later with the Israel Philharmonic [for DG in 1977], which we did in Berlin. Also I remember I wanted to perform the work myself conducting from the piano, and I said to Lenny: "Can that be done?" He said: "It's impossible. So go ahead and do it, Lukas!"

If you had to name the single most important influence on Bernstein as a composer, who would you choose?

Copland, I think. I would say Copland even more than Gershwin, although Gershwin is also highly important. But I think in the non-jazzy works you still feel Copland, and you don't feel Gershwin, whereas in the jazzy works the influences are probably about equal. Of course it would also depend on the nature of the project.

What do you think, in fifty or one hundred years time, we will remember Leonard Bernstein for?

Well, I think Lenny will mainly be remembered for his music, and *West Side Story* will eventually be understood for the deeply serious piece that it is. And I think at the same time his personality will still be around. With some composers it is not around at all. With some composers we know only the music. For instance, Bizet—who knows what Bizet was like? Whereas, with a figure like Beethoven, we all do know what Beethoven was like. Somehow with Beethoven the personality stuck. Another example is Copland. Copland's work will definitely be around in fifty years time, but probably with much less emphasis on the personality. Copland was a bit of a recluse, or certainly he was more of a recluse than Lenny! I think Lenny's personality will live on, along with his music.

NOTES

1. Bernstein in fact performed Copland's new Piano Sonata, not the *Variations.*

David Diamond

David Diamond, the American composer, studied composition with Roger Sessions and later Nadia Boulanger (1937–39 in Paris). Having received no fewer than three Guggenheim Fellowships, Diamond has had a prolific and varied career, which has made him one of the most respected composers of his generation. Diamond's symphonies have received their premieres from such major musical organizations as the New York Philharmonic, the Philadelphia Orchestra, and the Boston Symphony under such conductors as Koussevitzky, Mitropoulos, Bernstein, and Ormandy. Diamond had a friendship with Leonard Bernstein of some fifty years, their first meeting being in the 1930s in New York, while Bernstein was still at Harvard. According to Diamond, Copland and Marc Blitzstein "told me there was this extraordinary pianist there. . . ." Later Bernstein made his first professional recording with a Diamond composition, one of the Preludes and Fugues for the Piano, in 1940.

Was Bernstein's death a shock to you? You had known him since his student days.

The shock was more the suddenness of it, although I did anticipate it to some extent, and certainly a great many of his closest friends did; but we did not know it would be that way, that it would suddenly happen that fast. There are so many fables already, about what his last words were that I would just as soon forget about them. Bright Sheng [the young Chinese-

American composer] whom I know very well, was supposed to have been with him a day before he died. But I am not going to go into any of the usual highly romantic nonsense; I will say that it did shock me, in the sense that I am still not reconciled to the fact that that enormous genius—in the sense of genius within the human being, not he Leonard Bernstein as a genius, but the extraordinary quality, essence of life which is genius, and so rare—that that is gone forever.

Turning for a moment to Bernstein as composer, you have been quoted as follows: "The question I have always asked myself was why, with Lenny so gifted he didn't end up with Nadia Boulanger. She would have turned him into the most famous man in the world overnight." There seems to have been very little desire on Bernstein's part to study in Europe . . .

I think there we have something interesting about Lenny. Lenny had worked very hard with Walter Piston and Randall Thompson at Harvard [1935–39] and I must say, everything he was doing he knew somehow. When I heard the Clarinet and Piano Sonata [1942] he wrote for David Oppenheim, I told him I thought he had a magnificent sense of form. I think even with the conducting he always showed a knowledge of structure and instrumentation—he had had good training in that at Harvard. When I got back from Europe in 1939 after two years with Nadia I suggested his going to work with her, if not in composition then just in analysis for a few months—she was quite amazing with analysis. He couldn't have gone at that time because of his studies [conducting and piano] at Curtis and that coincided with the war breaking out. But later when Nadia was in America up at the Longy School I suggested it again and his response was that he thought it was a bit late. I told him with Boulanger it was never too late, that even Stravinsky went to her for advice. A lot of his friends, like Irving Fine and Harold Shapero were working with her, but maybe he felt he was beyond it.

What did you think of his early efforts as a composer?
I was very impressed with him—especially with *Jeremiah* [Bernstein's First Symphony]—he orchestrated a good deal of *Jeremiah* at my apartment on Bleecker Street. He wanted to get it into a competition [one sponsored by the New England Conservatory of Music, with Serge Koussevitzky on the panel] and he had only a few days to finish, so I told him to get over to my apartment and I would line up some paper for him. I checked some of the orchestration and, you know, he was a brilliant orchestrator. You can hear that in *Jeremiah.* As a first symphony from a young man this showed him to be a fine, solid craftsman. Nadia might have provided greater insight, but he had the ability to open up music in a way that was quite remarkable. He had a great success with *Jeremiah* when he first conducted it—in Pittsburgh—and I remember him coming back elated, it had gone so well. Then it didn't win the competition and he became rather depressed. But he eventually conducted the work all over the world and—a particular triumph for him—he got the Vienna Philharmonic to play it. He always said to me: "I'm going to show those Nazis." And he did; that was a great victory for him.

Would you say there was a falling off with the two later symphonies, Age of Anxiety *and* Kaddish*?*
Well, of course, in *Kaddish* it's Lenny being arrogant enough to think that he can have a conversation with God. That was Lenny, very typical of him. I know that [Dimitri] Mitropoulos was outraged. I don't like the musical attitude in *Kaddish* and I like the text even less—I don't like the work on either level. It ties in with his father to some extent. He was a Talmudic scholar, but at the same time an ignorant man and also arrogant. I remember he said to me the year after Lenny made his debut with the New York Philharmonic: "Do you think he's ever going to make any money?" I replied: "Mr. Bernstein, that's not what one thinks about."

But to return to the symphonies—from time to time when I hear *Age of Anxiety* I think it could work as an important piano and orchestra piece, but even with the revised finale I don't think it can quite make it in the way that the *Symphonic Variations* of César Franck or the Strauss *Burleske* do. Something is lacking there.

Could we move on to discuss Mass? *Bernstein was absolutely annihilated by the critics over this work. What were your feelings?*
There is the one work that Lenny and I really had a lot of disagreement about. I went to the premiere at the Kennedy Center [1971] and I hated the opening, with all the electronic sounds. I asked him why the use of electronic music was necessary—he had never liked this world—and then he told me: "It wasn't my idea!" He told me something about people involved in the production suggesting it. Lenny, you know, always had this need to be fashionable. What I disliked was the general showbusiness atmosphere—but what did impress me was the big aria and then the breaking of the sacred vessels—that to me is the masterful part and the only thing I really like in *Mass.* I thought *Mass* was his one big mistake, and a mistake of taste. But it was the 1960s when he conceived the work; everybody was doing their own thing and Lenny of course went along with it.

Robert Craft wrote after hearing Mass *that Bernstein suffered from a lack of identity as a composer, because of his "need to be all things to all people."*[1]
I think that is Craft's particular way of analyzing. Lenny having to be all things to all people is one thing, but I don't think this affected the music. This theory, this way of analyzing and deducting, I think is false—the man is not necessarily the music. I think Lenny was a compensatory musician; he knew his life was such a shambles, music had to be given every bit of the best in him. Of course neither Craft nor Stravinsky liked him—Stravinsky didn't like him at all, he didn't like the way he

conducted the *Sacre [Le Sacre du Printemps]*. Sometimes I would sit with Craft and Stravinsky and overhear them talking. Stravinsky was rather rude about him; he thought his carrying-on on the podium was pretty terrible and thought his music was awful. I remember his once saying to me: "You know, maybe he should be a Glazunov." Then when I asked him what he meant some time later he said: "Well, Glazunov was a heavy drinker too!" Of course Craft is something of an intellectual machine—that's why we have the false Stravinsky prose style in English.

Could we turn to the premiere of one of Bernstein's most autobiographical works, his opera A Quiet Place *(1983)? What did you feel about the work and what was Bernstein's reaction to the lukewarm audience response?*
I remember that in detail. When I got down to Houston I called Lenny at his hotel—I missed the dress rehearsal—and I asked him how he thought things were going. And he said: "I think it should work, but I'm not absolutely certain." Then at the performance the first thing that struck me, the thing I asked myself was how could Lenny, who was so astute and so good about the weaknesses in other composers' works and would often make wonderful suggestions for changing things—for the better—how could he have allowed that opening that went on for close to fifteen minutes of backstage choral music, with nothing happening, with the curtain down, with nobody ever being able to hear a word? That already turned the audience slightly off.

When it was all over and I was backstage, I told Lenny the work was very moving, but I suggested that he remove the prologue, that it really should begin immediately in the funeral parlor. [The first scene of *A Quiet Place* opens with a funeral, probably based on that of Bernstein's wife, Felicia, who died in 1978.][1] Lenny said to me: "There's something to it, but Wadsworth, you know, I can't get anywhere with that guy." [Stephen Wadsworth was the librettist for *A Quiet Place*.] I said to him:

"You're you. It's your opera. To hell with Wadsworth. You do what you want."

As for the Houston critics concerned, they were not very nice. I was standing with Michael Walsh, at that time I think already a critic with *Time* Magazine—I knew him from Rochester as a composition student—and Michael thought it was just terrible. I said: "Michael, you don't know what you're talking about; it's a very moving opera even if it does have all sorts of problems." He said: "It's got more than problems; Bernstein is not really a good, serious, musical composer." And Michael Walsh has always carried on that way, the way Harold Schonberg used to do in the old days, about the conducting as well as the music.

What happened was that the press was almost unanimous about the opera, and of course this did hurt Lenny terribly. He did not take criticism very well from the press; he suffered, but he suffered patiently, all those years of Virgil Thomson and Harold Schonberg flailing at him about his "chorybantic" behavior as I think Thomson once called it—about his inability to conduct quietly. How, Schonberg would say, could a man like Koussevitzky have encouraged Lenny to behave so flamboyantly on the podium—this is how they went on.

Schonberg was probably Bernstein's most severe detractor.

Oh, he was awful. I recall critical attacks even before the Philharmonic years. I remember Lenny's early appearances with the City Center Orchestra, often as pianist, playing Mozart and Ravel and Prokofiev—Lenny was a magnificent pianist. He was always trying to do as well as Mitropoulos used to do, because that was his model for playing and conducting at the same time. The more he did that the more Schonberg laid into him: why didn't he make up his mind whether he was a pianist or conductor. And then later when *Jeremiah* was performed: why didn't he make up his mind whether he was a better conductor or composer. Lenny would call me immediately. He had three real friends whom he could count on then: David Oppenheim the clarinetist, who was very devoted to him, and Marc Blitzstein and myself who were basically the people he would call for advice and suggestions—and Aaron [Copland] when Aaron was around. Aaron had already begun to move about, he was in South America a good deal, so he wasn't always available. Lenny always called us for advice in the early years and his complaints always were: "Why is this Schonberg after me, what does he want of me? Why is he giving so much attention in the press to me if he is going to attack me constantly?" I was appalled.

Schonberg thought nothing of discouraging Lenny; he regarded him as a show-off. Of course, there are people who still remember Lenny as being something of a show-off. He was; he was a flamboyant young man, because he was so sure of his talent. Mitropoulos told him once, gave him a lecture, asked him to be more modest, to be quieter, not to carry on so in public, but of course Lenny was not made that way.

Do you think Schonberg was aware of the damage he was doing, if not to the career then to Bernstein himself?

If Schonberg felt that at all, I would say it was minimal—the man strikes me even today as having a very thick hide. I don't think he feels a bit of remorse. I think Howard Taubman probably does, but Taubman was never as vicious as Schonberg was. I remember Taubman reviewing a performance of my Second Symphony that Lenny gave; he knew that the orchestra was not doing the best that it could, but he did not tear the work apart and he did not tear Lenny apart. A critic and writer like Deems Taylor whom I had a great deal of respect for—he was a famous critic in the twenties—he thought Bernstein was a wonderful talent. He and Taubman—these were the people that I trusted and who told me that this was without doubt a young man of genius. And I myself was never in any doubt, the moment Mitropoulos said to me: "You know, this is a genius boy; he will go far, but he will destroy himself."

Really, he predicted that from early on?
Oh, yes. Often when he came to visit me when I was living in Europe, in Florence, he would talk about Lenny. He would say: "You know, Lenny worries me. He has a very strange group of people around him, he's not sleeping, he has no discipline, he doesn't know how to take care of himself. He should be studying." I said that he studied all the time. And he would say: "No, you know Lenny, he knows everything backwards. So he studies at the last minute." Then when the Philharmonic problems started he [Mitropoulos] was eased out of his job and when he found out that Lenny had helped to knife him in the back, then of course, he did not care to see him very much. [Mitropoulos and Bernstein were co-directors of the New York Philharmonic in 1957, after which Bernstein was given the post in 1958.]

Bernstein and Mitropoulos seem to have had a complicated relationship. Bernstein, while competing with the older man, has always credited him as one of the most important influences on his own career.
Yes. He was Lenny's model for many things. When Mitropoulos fell dead at La Scala, Lenny was overwhelmed. Years later he would say to me: "That's the way I want to die. I want to fall right into the orchestra." And he persisted with this nonsense for years; and then when he began falling from podiums he said: "You see, you see—it's fated. I'm going to end the way Dimitri did." I told him not to forget that he would end with all the guilt within him too, that he bore toward Dimitri. I'm still waiting to get the crucifix back—that crucifix Dimitri had given to me. [Bernstein always wore two talismans when walking out on stage: Koussevitzky's cuff-links and a cross that had belonged to Mitropoulos]. Lenny took it away from me. He wanted it more than anything—he would do things like

that—he would just take them. I told him: "Have it." But now of course I want it returned.[2]

It seems that Mitropoulos saw seeds of a self-destructive tendency in Bernstein early on. One manifestation of this appears to have been Bernstein's incessant smoking . . .

Yes. I remember the cigarettes from as far back as Curtis, when Lenny was a student there. He always had a cigarette in his mouth and like everything with Lenny it was a competitive thing. I was a moderate smoker, smoked a brand called Half and Half and Lenny would tell me I wasn't a "real" smoker. Anything he could not do to the hilt, or anybody could match him at, was no good. Mitropoulos was worried about his inability to retain any form of discipline. You know with Mitropoulos—unlike Lenny—there was never any entourage whatsoever. I remember during the war years I would go with him [Mitropoulos] to a midnight movie on Forty-second Street or for a walk in Central Park during the blackout and I would take him to bars where soldiers and sailors came—he loved to watch them, he loved looking at men; but there was none of the frantic carrying-on that Lenny was doing, even in those years.

Could we move on to Bernstein's conducting?

Ah, well, there's the real thing. He had evolved a technique which began rather modestly in imitation of both Mitropoulos and Koussevitzky, with very little influence of Reiner whatsoever. I think that's why Reiner disliked him so, because the moment Lenny went to Tanglewood (Reiner after all had been his teacher at Curtis) he became more flamboyant, his gestures became bigger and bigger. Reiner had always told him not to use all that movement, that he was a wonderful talent, but that he would get more results, more volume, without it. Mitropoulos had always conducted rather flamboyantly, without a stick,[3] until the last years of his life, and Lenny took some of his wonderful visceral qualities. By the middle of the 1950s all these influences had produced pretty much what we know as the Bernstein style of conducting.

Some of the strange things that came later—for example, the holding of the baton with both hands—Lenny told me had nothing to do with the technical side of conducting. He would say: "My shoulders are in such vises, I'm pinned in the back; if I don't pull my arms forward, I'll never get through the rest of the work." That motion, of course, was thought by many people to be one of the craziest affectations they ever saw. Lenny as he grew older also had to deal with a certain amount of arthritis and rheumatism which must have caused him physical discomfort. But the technique—it *was* very personal, a very personal style of conducting—could be phenomenal and he knew how to get results. He was also remarkable with students and I think the fact that he knew he was aging badly was why he had begun to spend more time with youth orchestras, with young conductors. He came over to Juilliard on a couple of occasions and the students loved him. They would learn so much from him.

As a conductor he received enormous adulation from audiences; his following was at times more suited to the stars of popular culture than a classical conductor. Why do you think this was?

That I think was to some extent because of the Broadway shows. If Lenny had not been talented in that direction as well there might not have been the same great international reputation and the notoriety. Of course he would have been widely known as a very fine symphonic conductor but you see, many people knew him as the creator of *West Side Story, Candide,* and *On the Town.* Also, of course, there were things like the *Omnibus* series—he became a television personality—and he did fascinating musical analysis, it was a little showbizzy, but still convincing from the musical standpoint. He was always playing examples on the piano and then also demonstrating with the orchestra, and there was a commanding professionalism. He was a great musician at work in a medium that was new.

Bernstein was arguably the most famous musician of his time. Do you think he coped with his own celebrity?

Yes and no. It was the saddest thing. When I could still drag him away from the entourage, we would go say to Trader Vic's. We would go down the stairs and people would say: "Ah, look—there's Leonard Bernstein!" When he heard that he would look at me as if he could kill them. And I remember he said: "Who the fuck do they think they are?" And when they heard that, they turned away immediately, shocked. So there you have an example of the ambivalence in Lenny. In one sense he got to hate being stared at. Some people are able to handle celebrity quietly, discreetly, but Lenny didn't know what that was all about.

One elusive part of his career that probably troubled him more than any other seems to have been his lack of critical recognition as a serious composer. Would you say this was true?

Yes. That pained him greatly. Absolutely. He suffered because he was constantly attacked in the press. Even in Italy. I remember when he came for the premiere of the *Serenade* which Isaac [Stern] played in Venice. I went and I loved it—I still think the *Adagio* is a marvellous slow movement—but even there he told me the Italians had torn him apart. I told him this was common, it didn't only happen to him, it happened to Aaron [Copland], to me, to all of us. I asked him whether he had read Slonimsky's book [*A Lexicon of Musical Invective*]. He said: "Yes, but I don't have Slonimsky's sense of humor!" He suffered because he knew he was a fine musician, and he never understood why his serious music was not accepted. I told him if he were not Leonard Bernstein the conductor, and the composer of *West Side Story* and *On the Town* he might be taken in a different way.

To turn now to his last years, it has been suggested that after the death of Felicia in 1978, things became increasingly difficult for him. Do you see the 1980s as a particularly dark period for Bernstein?

Well, first, after Felicia died, he felt a great deal of guilt. I remember in the beginning when he brought me to meet her. She was like a beautiful boy, and he was knocked out by her. I never felt that Lenny was exclusively homosexual in the sense that he liked only men. I thought Felicia would be ideal for him in many ways, and I encouraged it, because I didn't like the way he was carrying on and the kind of boys he was carrying on with. There was this narcissism that went on with certain boys—he was always finding the type that he would have liked to look like. I felt that he would be better off with Felicia. But I told her: "You know this is not going to be easy. Lenny is strongly bi-sexual." But she thought all of that could be handled. And of course she didn't know what she was getting into. And later when he was not sleeping and life was becoming humiliating for her, she would say to me: "Is he always like this?" He had been in analysis for years, without telling me, and he helped me financially to go into analysis—this was one of the wonderfully generous sides of Lenny. And then we had the same doctor for years, Cyril Solomon, the man who arranged the commission for the *Chichester Psalms*. Chuck, as we called him, was reluctant to prescribe sleeping pills for Lenny's insomnia because with Lenny, once he took a sleeping pill, or once he smoked a cigarette, or once he took a drink—it was addiction.

Then later, particularly after Felicia's death I saw sloth set in, he was carrying a great deal of excessive weight, he ate like a pig, he would devour his food, and it was no longer possible to have a two-way conversation with him—he would never listen to anybody else talk. Toward the end he gave lectures, and sometimes they made sense and sometimes they didn't.

Do you think that in the last years of his life Bernstein was surrounded by too many people, that the entourage around him became damaging to him?
His life became overwhelmingly mobbed, by the sycophants and by that business world, that empire that was established around him. I have always felt that Lenny's beginning of his end was when the business enterprises began to develop into overwhelming proportions. I saw his life being taken over by his manager, Harry Kraut, and his henchmen—and a henchwoman by the name of Margaret Carson who was his public relations spokeswoman. Lenny had around him a kind of strange bunch; there were what I called the male harem, a group of very attractive young men who were always around him, and then there was the Kraut group, and of course Harry encouraged these young men to be around. That's where I felt the poison began and that's where I felt Lenny's life began to go down the drain. I always felt that if Harry had not encouraged these boys to be around all the time, Lenny's life would have been different—he would have gotten to bed on time, for example—but when they were around he couldn't bring himself to say: "Get going, leave, I've got to get some rest." There were a few people who were close enough to him who smelled immediately what was happening, who happened to know him when the career really began.

His behavior toward colleagues at this time often seems to have been harsh.
Yes. He treated many people badly at the end. In the last ten years his behavior could be shameful, to everyone for whom he should have shown the greatest respect. You see, the man had lost all judgment. There was the behavior at his daughter's wedding, in front of all the relatives, friends—in front of everybody. Lenny had had that one drink too many and when the speeches came he was one of the first to speak, as the father. He got up on the platform where the orchestra had been playing for dancing, and began to talk about his children and his son-in-

law. Talking about his son-in-law, he said: "You know what's so wonderful about him, he's not even GAY." And of course this threw everyone. [Zubin] Mehta, who was sitting at a table to the right of me said: "Can you believe it?" I said to him: "You heard." I looked over and saw Lenny's mother sitting there, and she was taking it as just another example of Lenny's carrying-on. Lenny was completely without shame. He was demoralized, de-everything. His self-destructive personality had taken over and it was obvious in everyone's eyes and it was painful for all of us, terribly painful. Everything was done to excess, there was self-indulgence in every possible way. What is extraordinary though is that he still had what he had from the early days—if he had to lock himself away to study before a concert, or when he had to prepare for a rehearsal, it was as before. But I could see the end coming because physically he had deteriorated so.

As a final question, what do you think is likely to last, first of Bernstein the composer, and second of Bernstein the conductor (the latter in terms of recordings)?

Well of course, that's a very difficult question because it's speculative, but if I had to answer as if I were playing a game or playing the horses I would say the *Jeremiah* Symphony and the *Serenade*. I think *West Side Story* may well be forgotten eventually, like so many Broadway musicals, with revivals from time to time; but I can't see *West Side Story* being that important in the next century—some of it will probably seem a little dated. *Age of Anxiety* I don't think will make it, in spite of the revised *Finale*, and a work like *Mass* will probably disappear totally. *Jeremiah* and the *Serenade* are what I would choose.

Now, as conductor, I think probably the best of Lenny went into Mahler. I am not a great Mahler fan; I like the Eighth Symphony most and after that the Ninth but I think Mahler may recede in twenty years' time. I think certain works of Stravinsky he conducted marvelously; *Les Noces* especially, nobody conducted this like Lenny, and even the *Sacre.* It amused me, although I never discussed this with him, that Stravinsky would pick on his tempi, because when you listen to Stravinsky's own early recordings of say the *Sacre,* and other works, they all vary. The *Symphony of Psalms* was the same thing. Also a work like Bartók's *Music for Strings, Percussion and Celesta*—that Lenny did wonderfully. I remember performances with the City Center Orchestra and they got better and better. But he was not a great Bartók enthusiast, he eventually got bored with the *Concerto for Orchestra.* Also I don't think he particularly liked the Ballet Suites [*The Miraculous Mandarin* and *The Wooden Prince*] although he liked the Violin Concertos. In some senses it is a pity that he did not spend more time on contemporary music in his later years—all his energies then were consumed by Mahler—but he had done a lot for contemporary composers when he was younger. Of course it meant a great deal to him as a Jew to have conducted and recorded Mahler with the Vienna Philharmonic; it was a personal victory for him.

NOTES

1. Robert Craft had written after the premiere of *Mass*: "The truth is that while Mr. Bernstein has a certain fragile personality as a composer—albeit difficult to uncover, his need to be all things to all people keeping him in a perpetual identity crisis—his resources as a composer are meagre. At any rate, he has not so far shown a very large command of a creative musical language."

2. David Diamond added the following when approving the interview: "Last year [1992], when I asked Charlie Harmon for it [the Mitropoulos cross] he wrote me it had been stolen with Lenny's jewelery when he was treated in the hospital. I have Charlie's letter about this."

3. Bernstein also dispensed with the baton for the first seventeen years of his conducting life. Later, after suffering from back problems while conducting in Israel in the 1950s, he took to using a baton.

II

Writers/Critics on Bernstein

Harold Schonberg

Harold Schonberg became chief critic of the *New York Times* in 1960, a post he held throughout Bernstein's tenure with the New York Philharmonic (excepting the 1958–59 season). He is the author of several acclaimed books, among them *The Great Conductors* and *The Great Pianists,* as well as *Facing the Music,* a collection of his Sunday essays. He has recently published a biography, *Horowitz—His Life and Music.*

The beginning of your time as senior music critic of the New York Times *virtually coincided with Bernstein's appointment at the New York Philharmonic. You have in some senses been his severest detractor. How did you view your role?*
Well, first of all, I never had any argument with Bernstein's talent; his talent was formidable—that was never in dispute. What I felt was that his ego was getting in the way of his music-making. There was a time, for many years, when he could not get a good review in a New York paper, in either the *Times* or the *[Herald] Tribune.*[1] But there was never any question about his natural gifts. You see, I think so much of what Lenny did was egocentric, whether as conductor or composer.

To turn to the compositions for a moment, you wrote after the first performance of Bernstein's Mass *that the work was "a combination of superficiality and pretentiousness." Do you think that* Mass *showed a questionable taste on Bernstein's part?*
I think *Mass* is an overblown, rather preposterous exercise in self-indulgence. In my opinion a lot of his music was pretentious—even *West Side Story* had this problem. Some of the earlier

musicals—*On the Town, Wonderful Town*—these had a light, witty quality. *Candide* also has some good passages—and the *Candide* Overture has worked its way into the symphonic repertory. But in *West Side Story* he was already trying to say something earth-shaking. And when he did that he usually ran into trouble. Lenny grappling with the infinite was a rather fearsome thing, I must say!

Do you mean by that the fact that West Side Story *was based on Shakespeare?*

Not only was *West Side Story* based on Shakespeare; Bernstein was, or pretended to be, terribly socially conscious, so he brought Shakespeare up to date, in a socially conscious atmosphere. I know many people think that *West Side Story* is the best of all his music theatre pieces. I find it cheap and sentimental. *Candide* is much more interesting.

I think he could have been the Offenbach of our time; there are many similarities. If you listen to the *Candide* Overture it has all the sparkle of Offenbach, of an Offenbach overture. An operetta or, rather, a musical like *Wonderful Town* is sophisticated and full of musical ideas, full of musical wit too as far as that goes. Of course he wasn't serious there.

To move on to Bernstein as conductor, you have written favorably about his early performances with the New York City Center Orchestra . . .

Yes, his work with the City Center Orchestra involved some very exciting concerts, and exciting repertoire. [Bernstein directed this organization from 1945–48.] I remember some vital performances of people like Stravinsky, Copland—there was an electricity at those concerts, and he had a young audience. Bartók, too—he did some very good Bartók. But then he took over the Philharmonic [in 1958], and everything changed. He was trying to impose his personality to such an extent that the music suffered.

You have described Bernstein in your book The Great Conductors *as being a throwback and a Romantic, particularly with regard to interpretive devices such as fluctuation of tempo. Could you elaborate on this?*
Yes. I think he overdid it, myself. Where most conductors of his generation wanted to be very tight—they all wanted to be Toscanini—he was much more in the school that started with Wagner, and went on to people like Bülow and Furtwängler, all of whom used extreme fluctuation of tempo. He was a throwback I think specifically to the Wagner and post-Wagner tradition. And he had a strong musical presence—I didn't happen to like it, but the presence always came through. Some of his performances of the Romantics, for example Schumann, I did like. I can't stand the Fourth Symphony myself so I never went if I could avoid it. But he did the *Adagio* of the Second Symphony very beautifully and of course it is one of the most beautiful things that Schumann ever wrote. He had things very controlled there; I was impressed.

Do you feel that with Bernstein some of his excesses with tempo fluctuation can be traced to his impulses as an educator?
Yes. He would underline things, make them clear, make them over-clear so that even the most stupid person could grasp what he was doing.

Do you think as he grew older he abandoned this?
Well, toward the end he had achieved "Old Master" status and he could do anything he wanted to. I think that by that time he had dispensed with any notion of educating the public; he was just reveling in what he considered the music to be, and his tempos got very slow and they were extremely personal. Sometimes he carried the thing off, because after all he was a damned good technician with a wonderful ear. But to me this was often against everything that the music stood for.

You have described him as a "perpetual Wunderkind" or "the Peter Pan of Music." Do you think he ever grew up?
No. No, I don't. You can't say though that he didn't live life to the full, in all of its aspects, both on and off the podium. Musically, I think he just got more eccentric and more personal. As I said, his tempos became slower and slower. There was a point when I could no longer bear to go to a Bernstein concert, in the last years. I remember a rehearsal at Carnegie Hall of Brahms's Fourth Symphony with the Vienna Philharmonic and the opening was so slow it was a travesty.

How do you view Bernstein's role as an educator, with Omnibus *and the* Young People's Concerts*?*
He was terribly charismatic, certainly. But with all of the talk about what he did, I still fail to see a nation of music lovers storming to get into concerts. Bernstein's television programs amused and entertained and I suppose in a way educated, but two hours after the program all was forgotten. Of the millions who listened to him, maybe a very tiny percentage would have said that he had perhaps opened their eyes to something. But the percentage had to be small. They've tried this sort of thing with children's concerts way back in the nineteenth century. I'm very doubtful—I think the only way you're going to get a music lover is if you grow up listening to good music. I had to review some children's concerts as a young critic and all the kids wanted to do was go to the toilet and throw spitballs at each other!

When Bernstein inherited the Philharmonic from Mitropoulos it seems to have been a very badly disciplined orchestra.
Yes, it was.

Bernstein didn't really change that. For example, he never managed to do in New York what Szell did in Cleveland.
No, but then Lenny wasn't interested in that kind of thing; he

was interested primarily in emoting. And the orchestra played for him—he was a good provider, with all the record contracts, television and so on. Mitropoulos was a great musician but he was so meek a man that the orchestra ran wild over him. It was better when Lenny took over. The orchestra respected him; they had reservations—I spoke to most of them over that period—but they decided to play for him.

Do you think the fact that he was American and not an import from abroad made any difference?

Well, the American thing might have been something, but let's face it, we're not talking about a man devoid of talent; there was a prodigious talent there.

You have mentioned that you think Bernstein had a wonderful ear and a clean-cut technique. Do you not think that the technique was at times so personal that it was difficult for musicians to follow?

Ah, but name me a conductor whose technique isn't personal. There's Karajan, with his eyes shut, waving his hands, or Reiner with his itty-bitty tiny beat, or somebody like William Steinberg, who's a damned good conductor, but I never could understand his beat.

What do you think were Bernstein's strengths in the repertoire?

Well, there was his Mahler, when he didn't go overboard. You know, the second movement of the *Resurrection* Symphony he pulled to pieces, trying to show how beautiful the melodies were, how folksy the melodies were. By and large he believed in Mahler and some of the things he did were very exciting. Also he was a very good Haydn conductor; I loved his strong, clear, bracing Haydn. He was very good at that. He was perfectly sound, given his egocentricities and given a certain amount of distortion that was inevitable with his kind of physical approach to the music. You certainly couldn't dismiss it. I never did—my own big argument was that his ego was getting in the way of his music making.

You have mentioned Bernstein's Haydn; it is interesting that with the other great composer of the classical period, Mozart, he very rarely seemed to have success. He didn't even program much Mozart.

But then who today would you consider a great Mozart conductor? What I'm driving at is that you can't blame him for something that's universal. And I think the authentic orchestras today are misrepresenting Mozart entirely.

To move on to Bernstein's serious compositions, do you see anything that is likely to endure?

I don't myself. But if fifty years from now he's pretty much in the repertory then of course I will have been dead wrong; frankly though, I don't see it.

How do you think the arrival of Boulez and Stockhausen affected Bernstein in the sixties?

I don't think the emergence of Boulez bothered Bernstein at all. It bothered Copland to feel left behind by the avant-garde—he started experimenting with twelve-tone then. I think Copland and Stravinsky were the ones who were unsettled by it.

In your book The Great Conductors *you appear to place Bernstein in a line following such figures as Jullien (a famous French showman-conductor of the nineteenth century) and Stokowski.*

Well, he was a showman, wasn't he? And in his day he was one of the three super-popular conductors—Karajan, Bernstein, and Solti. You can't argue with that. What is my opinion against so many? Compared with Bernstein, Karajan was rather motionless. For a man who was one of the great virtuoso conductors in history he would sit there very quietly, waving his hands. Of course, with Bernstein I think his physical appearance had something to do with making him a public idol. I wrote a piece for *Le Monde de la Musique* about him—an obituary piece—and one of the points I made was that his fairy godmother gave him all the gifts; can you imagine a bald Leonard Bernstein?

You've remarked that when he was with the (New York) City Center Orchestra he was doing some very good concerts; why, when he took over the Philharmonic in 1958 did you feel he was not able to continue that?
When he was with the City Center Orchestra, it was the most exciting, the most vital that I ever heard from him on the podium. When he took over the Philharmonic he was suddenly in the public eye, tremendously in the public eye, which he hadn't been before and he became a great deal more egocentric. But I don't want to get involved with amateur psychoanalysis.

Many commentators have suggested that Bernstein's real gifts belonged to Broadway rather than the concert hall.
Well, he never could stay with any one thing for very long. He wanted it all; conductor, composer, pianist, television personality, writing for Broadway. But yes, I think he could have been the American Offenbach.

NOTES

1. In the introduction to *Facing the Music* (a collection of Schonberg's Sunday essays) he has written that he believes Bernstein's unfavorable reception in the New York press had very little effect on his career: "Critics don't make careers. Artists make careers. A bad review in the *Times* may set a career back for a season or two. That is about all . . . . For years, as an example, Leonard Bernstein could not get a favorable review in the *Times* or the *Herald Tribune*. What difference did an unfavorable review make to him except bruise his ego?"

Joan Peyser

Joan Peyser studied music at Barnard College and Columbia University, and is the author of *Leonard Bernstein,* a biography (published in 1987), *Boulez: Composer, Conductor, Enigma,* and *Twentieth-Century Music: The Sense Behind the Sound.* From 1977 to 1984 she was the editor of the prestigious journal *The Music Quarterly.* She has recently published a biography of George Gershwin.

How does one define Bernstein? The character Junior in A Quiet Place *seems to be a fairly accurate portrait. (Junior is bi-sexual, has a traumatic relationship with his father, and is also subject to occasional psychotic episodes.)* Bernstein certainly based the fictional characters in *A Quiet Place* on his family, and Junior is, with some degree of creator's license, Bernstein himself. But I wouldn't call Bernstein psychotic; I think he is what the psychiatrists would probably call a normal neurotic, like most of the rest of us. But Bernstein has been allowed, because of the power that he accumulated throughout his life, to feel that he was entitled to act on everything. Most of us don't; we know that there are prices to pay. But with power and adulation and fame and money, he wanted everything (he wanted everything before he had all of that), both sides of every coin. He may have been homosexual but he also wanted a family life; he wanted to be respected and respectable so he had a conducting career, but meanwhile he played on and off with Broadway, which I feel was his real gift. Then there's the complex connection of his Jewishness. He wants to appear the committed Jew and then he ends up his conducting career as one of the primary conductors of the Vienna Philharmonic, which has got to be one of the most anti-Semitic orchestras in the world.

How about Bernstein, the Eliot Norton Professor of Poetry at Harvard, talking "jive" to a member of the Black Panther Movement? Would you say this was another example? (Bernstein's party for the Black Panthers was later immortalized in Tom Wolfe's Radical Chic and Mau-Mauing the Flak Catchers.*)*
Absolutely. And I think it's a little easier for Bernstein to talk "jive" to a Black Panther than it is to come through with the Eliot Norton. He did use people, one man in particular—Thomas Cothran—to help him with the lectures; he wasn't able to produce what he did by himself. Cothran was a very bright man; he also helped Bernstein with *Mass* which he was working on simultaneously. He made a tremendous contribution to the Norton Lectures. Whatever offended people about those lectures—the overall interpretation of them—was Bernstein's not Cothran's. The approach that tonality is God-given—this is what Bernstein set out to show—is an example.

To return to A Quiet Place *for a moment—musically the work is highly eclectic and even contains twelve-tone passages. Could you comment on this?*
Bernstein uses twelve-tone more to have an effect—of discordance, unhappiness, anguish—than anything else, and then he comes back to tonality. In other words it's the use of a language for his own special purposes, which in a sense is pejorative. When everything is in its place, then he uses tonality. People who embrace the twelve-tone language must regard this as a cheapening of that language.

A Quiet Place *was Bernstein's last major work and also his one attempt at "opera for the opera house"; its reception by both critics and public seems to have been a great disappointment to him.*
I think the biggest disappointment in his life has to have been the reaction to *A Quiet Place*. Only the opera companies who

were committed to it in advance of its composition presented the work. It was a tripartite commission (Houston Grand Opera, the Kennedy Center, and La Scala) and then the Vienna State Opera put it on as part of a deal calling for Bernstein to conduct some Wagner in exchange. Only those houses obliged to do it for one of these reasons did it. So it was performed not because it captured the attention or the love of audiences. My own feeling is that there was some quite beautiful music in *A Quiet Place,* and I feel that audiences and critics had a hard time getting past the libretto to hear the music, because they were so stunned by the repugnance of the characters. In the *New York Times,* critic Bernard Holland wrote that he was incredulous that a gifted man would choose these repulsive people to write a piece about. Nobody, absolutely nobody in the press read it as Bernstein's own family and his own life. I think he was very stung by that.

My own feeling is that Bernstein had been trying to tell the story of who he was since he left the New York Philharmonic. While he was with the Philharmonic he wasn't allowed to. He had to fulfil an image, he was watched every second. He had to conform. Once he left the Philharmonic—that and the death of his father occurred I think literally in the same week—he was really liberated. His father still had a big influence on him. He was a Talmudic scholar, a religious man, conservative and with Sam Bernstein's dying (I pronounce it "Bernsteen" because that's the way Sam always pronounced it) and the end of the Philharmonic post, he was free, and the first thing he did was *Mass.* He kept trying to confess from that point on and it didn't work. [In *Mass,* the key figure of the celebrant is in part an autobiographical portrait of the composer.[1]]

Then after *A Quiet Place,* I came along with the idea of writing something, and he told me: "Do it." There were probably some details that had he had access to a red pencil he might have deleted. But I think the man was much too smart, too clever, too powerful, to have encouraged me to the bitter end if he hadn't wanted a truthful portrait. And then of course he had to contend with all the acolytes around him who were screaming and saying, "Oh my God, look what she did." So by the end of his life I'm not sure whether he was as comfortable with the work as he had been when we first discussed it.[2] By then he must have wondered: "Well, maybe this wasn't so good for me." But it didn't damage his image in the public eye at all, or affect his career; if anything his career escalated in the last years.

This leads to quite an interesting point. Harold Schonberg has written (in Facing the Music *published in 1981) that Bernstein's inability in the first years with the Philharmonic to get a good review from either the* New York Times *or the* Herald Tribune *actually made no difference to his career, and that it merely dented his ego slightly.*

I don't agree. It may not have affected his career, because he was a terrific musician, but one of the things that I'm going to try to show in the Gershwin book [now published] is the incredibly lethal effect of public humiliation in the press. There's a big difference between what Harold Schonberg and Paul Henry Lang were doing in the *Times* and the *Tribune* in which they were ridiculing the jumping and everything and also the performances, and what I was saying in my book, which was: This man is a great conductor, he is a great composer of theatre music in the United States, he is a crazy guy, he is bigger than life, he is fiercely sexual, fiercely aggressive like all artists and this is the way that his aggression and sexuality show themselves—in other words I never diminished him as an artist. But they did, and that was not just denting an ego, and Harold knows that damn well—no one has a bigger, more fragile ego than someone in the kind of role that Bernstein was playing.

His son Alex said that his father would pick up the paper in the morning after a concert and say "Oh my God, let's see what he's done to me today." He dreaded it, naturally. We all dread being humiliated in public; why should he be any different? So I think that Harold was not fair—he and Lang transferred their own personal revulsion at his behavior into denigrating his gifts. And that was something else. I think it was very damaging. In fact, I'm not sure of this, but if there was one thing which I feel would have made Bernstein walk away from his discipline of the family life and the good father and good husband and so on, and move back to a more comfortable personal time for him, it would be the constant hammering in the press. He needed some nourishment, some support somewhere. I do believe it was so painful for him to see himself done in that way, every day, after a performance and certainly he needed some real pleasure in his life. I think for Schonberg to say that what he wrote in the *Times* has no effect on a person's career, and a person himself, is not correct.

Schonberg is not only highly critical of the conducting; he is also dismissive of a work like West Side Story.

Schonberg does the same thing on *Porgy and Bess*; like all of those people, also Virgil Thomson, even fifty years after the fact, they're still hammering away at it. It's an agenda: Somehow they feel that they have to keep classical music pure and away from anything that can be that pleasurable to watch and listen to; there should be some pain attached to art!

I think if I had to identify a trait that did not allow Bernstein really to fulfil his gifts, it would be the need to be respected and respectable. For example after *On the Town*, Koussevitzky came backstage and said: "If you do this anymore I'm not going to give you the Boston Symphony." So Bernstein stopped—this was in the mid-1940s—and all those collaborators, Comden and Green and Jerome Robbins, went with other people like Morton Gould—they did *Billion Dollar Baby*, they did a couple of other shows. Bernstein didn't do anything else on Broadway until Koussevitzky died, and the irony is that he didn't get the Boston Symphony post anyway.

He started again after Koussevitzky died, and he came up with *Wonderful Town, Candide, West Side Story.* Then he was offered the New York Philharmonic, and that also carried with it a prohibition against Broadway. So he didn't do any, because the Philharmonic meant respect and respectability. More than that, it meant big money and steady work, and Bernstein had a family to support, Felicia and two kids, who wanted to live well. Broadway seemed to offer a chancier existence, although after *West Side Story* I would have thought that he would have had confidence that he could make big money from musical theatre. But he went to the New York Philharmonic, and he was there for eleven years, until 1969 with no possibility of doing anything Broadway. So he again put the lid on his real gifts, which were in the popular idiom.

Some of George Gershwin's associates pushed him to stop Broadway and become an "artist." But he wasn't so gripped by categories. It's not as though he believed any of his songs would live—he didn't—he thought that they would die in two weeks and that only the large concert pieces might last a few years. Nevertheless fifty-four years later we still have "Love Is Here to Stay" and "They Can't Take That Away"—these songs are Schubertian in quality, so that if you don't pay attention to what the public tells you about categories and you just do what you are good at, that's the best way to make your life. And Bernstein didn't do that. I think we lost a lot of really important musical theatre that he was capable of doing, because of his need to be respectable.

You have made the following observation about his appointment at the Philharmonic: "As he moved away from entertainment and towards pontification, Bernstein's failures grew."

Absolutely. He became increasingly pompous and pretentious and he lost himself in the process. And then when he came back to himself it was to this silly, crazy guy. When I went on tour to promote the biography, I would tell people: However outrageous you believe the book is, it's not one-twentieth as outrageous as the man. I inhibited myself—I just wanted the essence of the man—and the guy was incredibly outrageous; that's not just normal ebullient behavior. It became even more so because of a feeling of resentment that he had to project this image that he knew was far from the whole story: the family man, the sweet man, the good man, the kind man, the loving man, and so on.

Something that seems to have troubled Bernstein, perhaps more than anything else, was the need to write a great symphonic or operatic work, to be recognized as a serious composer . . .

That's what I was saying about respectable. Somebody who was working in the international division of CBS records in the sixties when Bernstein was going to Europe to conduct told me that every time Bernstein was about to leave there would be a flood of memos from him [Bernstein] saying that under no circumstances was he ever to be identified as the composer of *West Side Story*. He felt that it pulled him down.

There was also the issue of the pronunciation of his name; the change from "Bernsteen" to "Bernstyne."

Yes. He made it what he thought was high-class, the German pronunciation, instead of low-class eastern European. The people who adopted the change with him were his brother and sister. Then when Sam, his father, died, his mother also became "Bernstyne." But Sam retained the old form; he knew who he was. He was a tough old man, an eastern European Jew and he was "Bernsteen" and he wouldn't have any truck with this other stuff. He also wouldn't be bamboozled by his son.

Of course, I feel that the connection between Bernstein and his father was the most important connection in Bernstein's life.

He obsessively replayed it, trying to make it come out right with every other relationship, for example, with Mitropoulos and Koussevitzky. In each case he had to one-up them, defeat them in some significant way. With Mitropoulos, he took a job away from him. And he did the same thing to Koussevitzky in a sense. Mrs. Rodzinski [widow of Artur Rodzinski, onetime conductor of the New York Philharmonic] told me that Koussevitzky went to the BSO a couple of years before he died and said: "You have to assure me that Lenny will become head of the Boston Symphony when I retire. Otherwise I leave." And they said good-bye. And he lost his job. And so the poor man did not die at the helm of the Boston Symphony. He died two years later, wandering around as a guest conductor. Bernstein was a destructive man in his relationships with others.

There seems to have been a need on Bernstein's part to annihilate father figures.
Absolutely. But he never killed off his old man. Sam may have been humiliated. The whole idea of his eldest son's homosexuality was very painful to him; he was an old-time, Talmudic believer and that was hard for him. But he survived. At the end of a rehearsal for *A Quiet Place* Bernstein was discussing the work with friends and he talked about the moment where Junior says to Sam (who is also "Sam" in the opera): "I love you, Daddy." Bernstein said that that was a tremendous spiritual orgasm for him and that every time he heard it he broke down and wept. So, finally, "I love you, Daddy" is the theme of Bernstein's life.

Could we discuss the symphonies? He seems to have made a good beginning with Jeremiah.
It was a good piece. He managed to combine in it various popular elements from his own life. [*Jeremiah* included the use of traditional Hebrew chants, as well as the jazzy rhythms Bernstein wrote into the *Scherzo*.] It even received praise from the critical establishment, which was unusual for a serious Bernstein work.

You have to remember he was very young, he had never done a major serious piece, he had just made his debut substituting for Bruno Walter that year, he was the big new figure on the horizon, and he was a kid, only twenty-five. The other factor which had an effect was that we were just learning about the Holocaust at that time and there was a bending over backwards on behalf of Jews, because people everywhere were feeling overwhelmed with guilt. So here comes this young Jew, he's a phenomenal conductor, and he has also written a good piece and everybody embraces it warmly. It's a little reminiscent of the positive reaction to Gershwin's *Rhapsody in Blue* a couple of decades before. But that was the last time most of the critics had anything good to say about Gershwin's work. After that, they hammered away in almost the same way, and his serious work was better than Bernstein's. The *Second Rhapsody, Cuban Overture, I Got Rhythm Variations,* especially *Porgy and Bess,* were pummeled by everybody. The attitude was: if you do it once or twice, we'll tolerate it, we'll even say it's good, but not after that; don't bring this jazzy, rotten, pop idiom into our hallowed halls.

Age of Anxiety *got some extremely bad reviews—"a masterpiece of superficiality" from Olin Downes—but then it doesn't seem to have been as good a work as* Jeremiah.

Generally with Bernstein's compositions I feel that virtually every melodic idea is borrowed, almost intact, not even transformed, from other people; I don't mean the kinetic, athletic, jazzy stuff but the melodies. He was not a melodist. He himself knew that. He once wrote a little article entitled something like: "Why don't you go upstairs and write a nice Gershwin tune?" People would say that to him, but he understood how hard it was to do that. He did not have that gift. Of course the reviews got worse and worse. And with *Kaddish* people were very offended by his conversational tone with God. The relationship with the father is not only mirrored with Mitropoulos and

Koussevitzky but also with God! Particularly in *Kaddish,* he sort of lectures God.

And in Mass *he tries to challenge another God?*
Exactly.

Could we talk about the arrival of Boulez and Stockhausen on the musical scene, and how the avant-garde movement in the sixties affected Bernstein?
There was a point when everybody was attacking Bernstein because he was such a conservative figure and was writing and conducting only tonal music. Schonberg was attacking him on this even though he (Schonberg) hated post-Schoenberg work; whatever these guys did they couldn't win. Then Bernstein did program several months of new music (meaning the post-Schoenberg—post-Webern line), but what he did was crazy—he would apologize for it to the audience. Gunther Schuller remembered his apologizing in advance to an audience for doing a Stockhausen work. He would say: "I know you're going to hate this, but you've got to listen to it, it's the right thing to do."

His ideas were obviously rooted in tonality, and in a tonality we haven't seen since maybe Brahms or Wagner. That of course was the purpose of the Norton Lectures—to show why he was right about tonality all this time. So when he performed twelve-tone music, it was often with an effort to show how bad it was. I remember interviewing Milton Babbitt about one of his pieces that Bernstein was doing with the Philharmonic in 1969. The rehearsal sounded like a nightmare. Babbitt said that Bernstein didn't seem to know what he was doing and that he kept apologizing for the work in front of everybody. But Bernstein was obliged in a sense to move with the times, and he always wanted to be seen as *au courant,* to be thought of as avant-garde, even by incorporating some twelve-tone material into his own works. But his heart was in tonality, and that's why I feel he would have done better to have stayed in the popular music field where nobody has ever really questioned that. But he could never commit himself in any one particular direction.

Ned Rorem, after Bernstein's death, referring to the criticism that Bernstein had "spread himself too thinly" said that his very identity was as a "Jack of All Trades."

That's true. In one sense one can only try to explain a life, one cannot judge a person and say he should have done this or that. Bernstein's life was very rich; he probably got out of it most of what he wanted. But there was this tremendous disappointment with *A Quiet Place* and also I know that when he looked back on his list of concert works he felt disappointment at what he had produced and what was likely to have lasting value. Basically he was a collaborative musician and what is interesting about him is his passivity. He may have looked like this blustery, aggressive personality but if somebody said: "Lenny, take out these six measures and put them over here," he would say: "Do it." He was very easy to work with, everyone who was involved in the Broadway collaborations says that. And *A Quiet Place* was reorganized by other people; someone would try putting *Trouble in Tahiti* [an early work that Bernstein incorporated into *A Quiet Place*] in the middle, someone would take it out, someone would try it at the beginning, and so on.

Could we talk about his conducting? The choreography is of course well known—what do you think prompted this?

I think it was the theatrical nature of his personality. In high school and later at Harvard, according to Mildred Spiegel, who was a platonic girlfriend, he would go to parties and play the piano standing up with people crowding around and then he would move on to another party and everybody would follow him like the Pied Piper. She spoke about this incredible electricity. He hungered for adulation, he had a need for this constantly. That seemed to be his basic temperament. People remember him even as a kid in school generating excitement and that is something that one can't discuss in a causal way. It happened so young. And naturally it translates into his athletic conducting style. What he does on the street with friends, he does on the podium.

He saw that it worked. Look at what happened with his debut with the Philharmonic: he made the front page of the *New York Times.* So he wasn't going to throw all that away. There's a story of how, when the criticism got really severe, he inhibited himself and conducted the way anybody else might have, I think a Brahms work, and afterwards the musicians came up to him and said: "My God, Lenny—what's the matter, are you sick?" And so he knew that that hadn't worked. I think he felt that the choreography was not only useful for his public image, but also for the musicians, that it whipped them into some kind of frenzy. If you watched him in a rehearsal, he was not like that; he was all business. He would sit on his stool and it was all hard work. And of course he talked too much, he spent much too much time interpreting things vocally.

And sometimes talking about completely extra-musical topics!
Very often! So that he wasted some rehearsal time. But then when it came to the performance he did this thing, and he thought it made a real difference in the performance, and I think he was right. I think that contributed to the electricity. He asked a friend, Arthur Bloom, whether he should conduct the reconstituted version of Mahler's Tenth Symphony, and when Bloom told him it was for him to decide, he said: "I have one criterion. Will it give me an orgasm?" He had this incredible sexual energy. One surprising reaction to what I wrote was that describing his sexuality the way I did was considered pejorative; I regard it as life-enhancing. And the orchestras wanted to play well for him, as exasperating as he was. And he made the Philharmonic richer than it had ever been, with the television shows—*Omnibus,* the *Young People's Concerts*—all the recordings.

To move on, Richard Burton, the British actor, wrote in his journal for 1980 as follows: "Bernstein is indeed a fascinating creature, genius and dolt, a man and a woman. A boy and a girl. There is no personal hell quite like the hell Lenny lives through. All the time, night and day there is the battle between his super ego and his utter self-loathing—a Mahatma Miserable. I think that master means to die shortly unless the will to live reasserts itself. And all those faceless sycophants around him. Repulsive." Do you think the last ten years were a particularly dark period for Bernstein?

I think his whole life was a dark period for him. I think he thought that he would have freedom after 1969 [the year Bernstein left the New York Philharmonic] and that freedom would mean all choices were possible. So he went for everything, and it was disastrous. Burton is certainly correct about the self-loathing. He was a self-loathing man. I think that after 1980 he got some juice to live because he was working on *A Quiet Place* and then that failed. As for the sycophants, he not only tolerated them, he encouraged them. I remember talking to him at Tanglewood after a rehearsal of Beethoven's Seventh. There were five or six young men around him in the green room, and Bernstein was taking his shirt off, preening and so on, in his usual way. He asked me how much time I wanted and I remember I said I would rather have five minutes with him alone than ten hours with all these people. And he turned around, lifted his hands in this charming gesture and said: "Boys, she wants me alone!" You see, he usually played to that audience—of those boys—and then you didn't get a real person. And he had that entourage with him everywhere. Burton is absolutely correct.

It is interesting that, for example, Luchino Visconti was also very often surrounded by a group of acolytes, especially in later years.

Visconti and Bernstein worked together in 1954 I think, in Italy with Callas. And she made some comment about all the attractive men in the world being homosexuals. It is fascinating that after all those years Visconti and Bernstein ended up in much the same social environment. And of course it's very destructive, but there must have been something in both of them that needed it. By 1980 Bernstein wasn't responsible to anybody; Felicia had died, his children were all grown up. And at the end he looked bad you know, ravaged.

Do you see him as a type of "Dorian Gray" figure?
It is a Dorian Gray type situation, because was there ever a handsomer young man? He was a most beautiful young man.

His treatment of some friends and colleagues in the last years seems to have been rather bad; an example is his behavior toward Maximilian Schell during the making of the Unitel films on Beethoven. Is there any explanation for something like this?
He couldn't bear anybody else to have center stage; he had to be the star all the time. Anything anybody else had he had to have, he had to have the food off everybody's plate, he had to humiliate everyone who was in touch with him, he even took lovers away from people who were close to him. His behavior to other people could be abominable. I don't think a person who is not a self-loather would behave that way.

The man who worked on the Unitel films with Bernstein and Schell, Bill Fertik, said that his whole career was inspired by Bernstein's Omnibus *shows. Could we talk about Bernstein's role as an educator?*
I have found many people who have said that they went into music or something related, because of the early *Omnibus* or the *Young People's Concerts.* I believe these people. What I feel may not be true is that whole generations grew to love classical music because of Bernstein. We don't have any evidence that that is true today. Few people are interested in classical music today, and these are people who were kids then. Business is very bad for art music, certainly in the United States; it may not be suffering quite so much in Europe. What I am saying is that unless the person was initially driven in that direction, what one

came away with from—say—*Omnibus* was probably more an awe and appreciation of Bernstein than of classical music per se. But if you were gifted musically, he did turn you on. I do think his shows were wonderful. He was always sophisticated but not pompous in his presentation. Also, it was the first time anybody had done anything like that. There has always been this terrible thing called music appreciation, which means you put it up on the mantel and genuflect. He never did that. When I say I don't think he converted whole generations into lovers of classical music, that should not suggest that he could have done it any better. I don't think you can convert people in that way, in this society today with so many diversions and distractions. It's an uphill thing to try to get large groups of people to love art music.

What do you see as Bernstein's legacy?
That's very hard. It's as hard to predict that as to predict the stock market. All I can say is what I feel about the stuff today. In terms of the music I think *West Side Story* certainly has lasting qualities. At the present time, it is probably being performed somewhere in the world every day, whether on film or live. The *Candide* overture is also I think important—it is the most original piece of music he wrote. As a serious composer, I don't think his future looks so good. I think that's what he knew and what disappointed him so much at the end. I think he felt that he had not left the legacy that he had perhaps been capable of, had he lived his life a different way. It may sound presumptious but I really don't think you can count on a long life for works like *Age of Anxiety* or *Mass*. Also the other thing that Bernstein must have been conscious of—he was in many ways a self-aware person—was that very few conductors other than he ever programed his works. Since his death we haven't seen everybody running to record Bernstein. I think the theatre music is what will continue to be presented—that seems to me to have a long life; I hope so.

As a conductor, on records, I find his Haydn and Stravinsky particularly wonderful and I think at the end when he got deeper and deeper into German Romanticism, a lot of those performances were very fine. At the beginning of his career he was very good to American composers—he used to do David Diamond and William Schuman, for example. But by the end there was some resentment because he wasn't as committed. By the end of his life he did just what he wanted to do, and nineteenth-century German music was what he loved. It's interesting to compare the two complete Beethoven symphony series—the ones he did at CBS and the ones at Deutsche Grammophon. Each of them has a different kind of value, but they're both positive. I think the recordings will probably last, but how long I don't know. After all we don't sit around now listening to Toscanini all the time.

NOTES

1. Joan Peyser has written in her biography of Bernstein: "Through both music and text [in *Mass*] Bernstein traces the course of his life from the Greek, that is the Mitropoulos influences . . . through the composition of Broadway shows, to the inclusion of a kaddish. There can be no explanation for a kaddish in a mass unless *Mass* is in fact the story of Bernstein's own life."

2. According to Humphrey Burton's recent biography, Bernstein never read Peyser's book.

III
On the Record

Paul Myers

Paul Myers worked as a producer for Columbia Records (later CBS, and now Sony Classical) for some twenty years, making records with such illustrious artists as George Szell, Vladimir Horowitz, Murray Perahia, Isaac Stern, Pierre Boulez, and Glenn Gould. As vice president of Artists and Repertoire for CBS in Europe, he was closely involved in the discussions that preceded Bernstein's departure from the company in the mid-seventies. Paul Myers is currently one of Decca's leading producers, based in London.

Could we begin by discussing Bernstein's break with CBS and his move to Deutsche Grammophon? The negotiations to release Bernstein (from CBS) appear to have been rather fraught . . .

First of all Bernstein was not CBS's leading classical artist in terms of sales. This may surprise people, but in fact that position was occupied until 1968 by Eugene Ormandy and the Philadelphia Orchestra.

I remember when CBS sent me to London in 1968, I couldn't understand why Bernstein was not a major name. I remember saying to a colleague [Ernest Fleischmann]: "But Good Heavens—he's the man who wrote *West Side Story* . . . " And the reply was: "Yes, we try to keep quiet about that!" There was still a great deal of snobbery surrounding classical music at that time.

CBS had always been a hardline selling company. They had made enormous sales of *West Side Story* and one particular Bernstein record of *Rhapsody in Blue* and *An American in Paris.* Bernstein's other recordings did sell well, but he was always being sold to the American public as a Gershwin man and as a Broadway composer. Whereas, of course, he was also a major conductor. To be fair to CBS they had also done the first complete Mahler cycle with him. But American recordings were not popular in Europe in the late sixties/early seventies because the sound was somewhat different and rather bright and there was a lot of what the magazines call spotlighting. You could say that in the sixties European recordings always sounded about ten feet too far back, and American recordings sounded about ten feet too far forward. Over the next decade or so, the Americans moved back five feet and the Europeans moved forward five feet and so we got a kind of universal sound!

I was situated in London as vice president of Artists and Repertoire by the mid-seventies. Bernstein wanted to go. Harry Kraut was very eager to see Bernstein develop a more important European image and Bernstein wanted to strengthen his ties to the Vienna Philharmonic. Deutsche Grammophon believed that signing Bernstein would help them to conquer the American market. DG had never really achieved a major breakthrough in America. It dominated the *European* market and was certainly extremely influential in Japan, but not in the States.

Frankly, Bernstein and CBS were getting a little bored with each other. Goddard Lieberson, the president of CBS, had signed Bernstein for a twenty-year contract because he had so much faith in him. Bernstein had *carte blanche*—he could record anything he wanted to and we had done almost his entire repertoire. But CBS was not selling him as it had done in the sixties. I looked at sales figures and in fact these were not good—it had reached a point where many of his records were not breaking even.

Bernstein's relations with CBS seem to have been on a plateau at this time. Would you say that his career was as well?

Perhaps. Bernstein was looking more and more toward Europe, and he was coming to the end of his tenure with the New York Philharmonic. It was like a long-time marriage in which familiarity had bred not contempt but a certain over-familiarity, and a new Bernstein record on top of the two hundred and fifty odd he had already made was not generating any excitement. I said to the marketing people: "Look, we are not selling Bernstein anymore. We've recorded just about everything—why are you hanging on to him so hard if he wants to go to DG?" Also, a rather sensitive situation, we were not entirely convinced by the man who was doing his production, John McClure. It's difficult to say whether he was right or wrong for Bernstein. He certainly kept Bernstein happy, which is one of the important jobs of a producer, but there was a certain amount of dissension within CBS as to how good his production was. But whether that was Bernstein or McClure is very hard to say. Anyhow, we let Bernstein go.

Moving back to Bernstein's early years, do you think that as a conductor he resented his Broadway image?

Absolutely. Quite correctly, he bitterly resented being identified simply as a Broadway whiz kid. He was one of the great musicians of this century, as far as I'm concerned. He made the public aware of Mahler, and that probably more than any other conductor. He inspired at least two generations of young American musicians. As a communicator about music and—his own title—the *joy* of music, there is nobody in this century who has had that power. His television programs, on Beethoven and on all sorts of musical subjects, were magnificent. We're talking about a man to whom you have to attach the word "genius," even though it is one I hate seeing overused. "Genius" is like "unique", it is not comparative and you have to then say: Does he join the ranks of Michelangelo and Einstein and so on? Perhaps he doesn't, but as a man of enormous brilliance, communicative ideas, appeal, and everything else, Bernstein was a major musical figure.

Of course, for many Europeans, his excessive style of conducting was slightly embarrassing; they weren't used to seeing this man leaping about and carrying on. As he said himself: "I'm the only man I know who is paid to have a fit in public."

It was a difficult hurdle for him; he had been the great Broadway idol, who had made the transition from Broadway to the concert platform. That's very difficult to do, and who else has done it? Yes, Dr. Eugen Blau, who many years earlier conducted the Roxy Theatre Orchestra and later changed his name to Eugene Ormandy! Of course, one or two new conductors are doing it steadily. John Mauceri, a Bernstein protégé, is one example. But in those days it took much longer. Then if you were a classical person you took pop records home in a brown unmarked envelope, and if you were a pop person you didn't tell your friends about the secret vice of listening to classical music! Today, of course, taste is much more catholic, in every way.

In the spring of 1968, Bernstein made his famous recording for CBS of Der Rosenkavalier *at the Vienna State Opera. Could you comment on how he approached the work?*

The sessions were very interesting. I would say overall the thing that Bernstein always enjoyed was to be in the center of a drama, which seemed to be falling apart, and then at the last moment he would save the day. It was *almost* deliberate. Not quite. I think he enjoyed playing the part of the hero who rides in with the cavalry at the end. He wasn't entirely happy conducting *Rosenkavalier,* and the dress rehearsal didn't go terribly well and he was wise enough and astute enough to recognize this. I think he disliked the fact that he had already reached the "Maestro" stage where nobody would tell him the truth. I remember going backstage after the dress rehearsal with a friend who said to him: "Well, maybe it will go a bit better on the first night." And

Bernstein's face lit up and he said: "At last somebody who will tell me the truth." He was of course accustomed to being surrounded by a sycophantic set of admirers and I think this did worry him at times.

To move on to your personal relations with Bernstein—what was he like? Was he easy to work with?

I knew him on a professional level as the director of CBS Masterworks for Europe and he had known me as a young producer hanging around his sessions. Later I had the official job to greet him when he came to London for concerts or recordings. He was always very pleasant, polite, extremely nice to me. I had no complaints whatever. But when he drank a lot, which he did, he used to become my best friend and I felt a little bit like Charlie Chaplin in *City Lights* with the millionaire who doesn't even know him when he's sober and regards him as a long-lost friend when he's drunk! On a number of occasions, when he was in his cups, he used to weep on my shoulder.

There was always the suggestion—very often put about by Bernstein himself—that he had not fulfilled his early potential as a composer. Was this something you felt?

Oh yes. I think the burning ambition of Bernstein's life was to write a serious masterpiece, and it worried him enormously that he didn't. I hate to say that about him now, because I admire his music enormously, but I don't think he ever *did* write a masterpiece. Except for *West Side Story*. That's a masterpiece, in its own genre.

I can remember a party we threw for him on his fifty-fifth birthday at the Edinburgh Festival. He literally put his head on my shoulder and cried. He said: "I'm fifty-five, two years younger than Beethoven was when he died, and I haven't yet written a masterpiece." It was almost as though he felt that the world was waiting for him to write this masterpiece and he didn't have the time to do it. Anyway, a masterpiece is something that just comes out of you; you can't sit down and consciously write one. And I think he felt that that was what should happen.

As he got older there was a certain amount of what you can only call self indulgence about his life style. He didn't dedicate himself to his work. He didn't live only to work. At least not in the years I knew him. He was gregarious, he was wonderful company, and he loved his Broadway friends. I can remember standing in a cinema queue in New York at the Sutton, seeing Lenny with Adolph Green and Betty Comden, waiting to go in. He loved to be a man about town—maybe he was a Rossini reborn, but he wasn't a Beethoven reborn.

He had this fixation about Beethoven. Humphrey Burton made a Beethoven bicentenary film with Bernstein. It was in three parts and in the first part Bernstein, at his most brilliant, talked about Beethoven and the logic of Beethoven and how a man whose domestic life was totally chaotic, totally in shreds, could write music that was the most logical ever written. In the second part of the film Bernstein conducted sections from *Fidelio,* and in the third part, before conducting the last movement of the Choral Symphony, Bernstein said he would like to do an off-the-cuff little speech. What he said was more or less as follows: "I am standing here in the pit of the orchestra, where he [Beethoven] stood. He was a pianist and I am a pianist. He was a composer and I am a composer. His initials were L.B. and so are mine . . . ." It was frankly embarrasing.

Do you think this sort of remark stemmed from some kind of insecurity on his part?

Perhaps. I remember Lenny telling me something similar about Mozart. He told me after he had played Mozart's piano, and many, many musicians have done this, that he felt that now he alone really understood Mozart's music!

To move on to Bernstein as interpreter, many people have found his music-making too personal and egocentric. Was this something that you felt?

Well, I remember a concert where he conducted Elgar's *Enigma Variations* in London in the 1980s. It was a *very* odd performance. It went on literally for ever. It was nearly forty-five minutes long and *Enigma* is usually thirty-two or thirty-three. I went backstage and spoke to the first violin, Rodney Friend, and said: "Well, that was an *enigmatic* performance!" And he said: "If it had gone any slower, it would have gone backwards!"

Then I went to pay my respects to Bernstein, and he was pronouncing, really quite arrogantly: "Of course, none of these British conductors understand Elgar the way I do. All these Boults and people—they didn't understand what Elgar was about!" And you never knew whether he was covering up some inadequacy or whether he genuinely believed it.

Do you think Bernstein's more whimsical tempi came from a need to be different, a need to put his own indelible stamp on a work? What do you think prompted them?

I really believe he was convinced by what he did. The closest parallel I can give you is perhaps Glenn Gould. I was a great friend of his for some twenty years, and Glenn never did anything capriciously. He did what he believed in, and he believed in making music that way, and I think Bernstein did too. I think he became totally immersed in the music, and you therefore got his personal reaction. But there was a difference between Glenn and Lenny: for example, Glenn once was interviewed on the Beethoven Sonatas and he said: "I'm interested in Beethoven because I'm interested in the structure of composers, but if you want to hear Beethoven Sonatas played as they should be, listen to Schnabel, don't listen to me." He was really quite modest. He regarded himself as eccentric in what he chose to play, and hoped people would enjoy it. Bernstein was more inclined to say: "I'm playing it because God told me this is how it should be."

The strange thing is that orchestras, for example, the Vienna Philharmonic, would accept this kind of pronouncement from Bernstein . . .

Yes. But you must remember, this was a man of *tremendous* personality and imagination and magnetism. Lenny was a most endearing person. And I think as musicians the Vienna Philharmonic or the New York Philharmonic respected him. One of the problems with the London orchestras is that they play with so many different conductors and have so little rehearsal time, that they are inclined to be a bit offhand with all conductors. Don't forget Bernstein was the only man who ever tamed the New York Philharmonic, which is famous for being the toughest orchestra in the world. I have seen conductors with tears in their eyes trying to control that orchestra. The New York Phil. loved Lenny. He was one of them. He brought them enormous fame, he brought them lots of money from recordings and they adored him. He single-handedly took the slightly po-faced attitude of American orchestras—and audiences—toward classical music, and changed them, brought them into the twentieth century.

To move on to Bernstein's role as an accompanist on the podium, he does not seem to have been particularly sympathetic. When Murray Perahia performed and recorded the Schumann Concerto with him in the seventies, Perahia appears to have been unhappy with the result. He even asked that the recording not be issued . . .

I remember all of that very well. Early on in Murray Perahia's career I had produced two Chopin Sonatas for him and he then went to the New York Philharmonic to do the Schumann Concerto with Lenny. Murray told me that at the rehearsal Lenny didn't have time to discuss anything, so he asked him if he could discuss a few points afterwards. Lenny said: "Oh, well, yes, we'll do it at the concert."

For some reason the Schumann was in the second half of the concert and so at interval Murray went, hat in hand, to Lenny's dressing room. Lenny had some dancer friend there and ignored

Murray, refusing to talk to him. It was only when they were going down in the elevator to the stage and Murray said: "Look, could we . . . " Lenny said: "Oh, don't worry, I'll just do whatever you do." And that was the entire discussion.

Then the recording took place. And Lenny was at his worst—everything was totally over the top. Murray, with tears in his eyes, played me the tape and said: "Please don't release it, I'm prepared to pay for the cost of the sessions." We didn't ask him to do that and I made the decision not to issue it. (It cost fifty grand.) Lenny was very offended and we found some suitable words to say that Murray was very unhappy with his own performance. If you were ever to hear that tape, you will understand what a travesty it was. I wasn't that much in favor of Murray doing the recording with Lenny anyway. Murray is a great Romantic Pianist (with a Capital R and a Capital P) and he is that mixture of Classicism and early Romanticism which is what Schumann is all about. It is not something which can be performed in a sentimental way.[1]

Another difficult meeting seems to have taken place with the Glenn Gould/Bernstein Brahms First. Do you remember these performances?

Yes. It was a total storm in a teacup. The person who really caused the problem was Harold Schonberg. It was very naughty. Schonberg's "Ossip" letter was totally unfair. [In his review, Harold Schonberg wrote one of his "Dear Ossip" letters. Ossip Gabrilowitsch was a famous pianist-conductor from the earlier part of this century. The letter went as follows:

> I mean this Ossip. Glenn Gould is waiting in the wings . . . and has to listen to Bernstein saying that this was a Brahms he never dreamed of. He washes his hand of it. He says, believe me, Ossip, the discrepancy between what he thinks of the concerto and what this Gould boy thinks of the concerto is so great that he must make clear this disclaimer . . . So then the Gould boy

comes out, and you know what, Ossip? . . . . He played the Brahms D Minor Concerto slower than the way we used to practice it. (And between you, me, and the corner lamp post, Ossip, maybe the reason he plays it so slow is maybe his technique is not so good.)]

I was with Bernstein the following day at a recording session and he was horrified by what Schonberg had written. The whole point of his pre-concert speech was to say: "The great joy of music is that you can have two completely opposing views and they're both right." So the remarks by Schonberg that Bernstein had washed his hands of the whole thing (and that Glenn didn't have enough technique) were most unfair. Certainly Bernstein never intended to harm Glenn. He adored him, and they got along very well. Glenn worried about him. I can remember his saying: "You know, Bernstein is the eternal Peter Pan. He's never going to grow up and he's going to be faced when he gets to the age of sixty by terrible crises in his life. He can't ever accept the fact that he is no longer young." And I think he was probably right.

Two things happened. The death of Felicia I think meant a lot to him. And I suppose you have to talk about the Jewish personality and the sense of guilt. I went to see Bernstein at a concert he had conducted in Paris. He had grown a beard, which, incidentally, Harry Kraut later made him shave off because he had too many pictures of him without one! I went backstage and said: "Lenny, how are you?" He said: "Oh, I'm wonderful. And it's so great to feel free."

At this point he was living with a young man[2] in Los Angeles, and described how they would sit in the garden at night reading Tolstoy aloud to each other. Then he learned that Felicia was dying of cancer, and went back to her and felt terribly guilty about the whole thing. I didn't see him for a couple of years and I was shocked by the way he had suddenly aged. In the last few years of his life it was almost like Dorian Gray's picture. All of his excesses seemed to come out in his face. Also I had remembered him with that very slim, athletic body and suddenly he was shaped like a pear. He became quite distended and it was a real shock.

Do you think that by this point Bernstein had surrounded himself with too many people, too many sycophants?
I would say with him that there was a certain cynicism with personal relationships, because that was his lifestyle and because also he learned to mistrust some of the people around him. I think there was a constant tug of war between needing people around him, to encourage and adore him, and at the same time realizing that some of those people, because of his star status and his *power* in the music world, would say anything to please him. It's very hard because all my views are only speculative; I felt that he was supposed to be the *Wunderkind* of all time, and, after a while, he was having trouble living up to his own image.

Earlier, when Schuyler Chapin was his manager, Bernstein's life and career seem to have been more under control. Do you think Chapin had a beneficial influence on him?
Yes, I think when Schuyler was around things were probably better. [Chapin preceded Harry Kraut in running Amberson Enterprises.] Schuyler made Lenny behave. He would say things like: "Lenny, we don't do that." And Lenny would listen to him. Apart from anything else and this may sound snobbish, Schuyler came from a very good old family and Lenny with his Jewish Brookline background from Boston had a kind of respect for people with inherited class. In this day and age it sounds a strange thing to be talking about, but it's true. Lenny listened to Schuyler. He also listened to Felicia when she was around.

Felicia was very understanding. She always said: "I would sooner have a half of Lenny than the whole of anyone else." When they came to London, Felicia would rather discreetly leave at the end of the recording or whatever, and she would go back to New York while Lenny and Harry would go off to Rome to play.

I was in Rome for the Vatican concert. There was a wonderful story that somebody from the Bernstein entourage had approached one of the Cardinals and said that Mr. Bernstein would like an audience with the Pope. The Cardinal said: "I'm sorry, but His Holiness has got rather a busy week . . . " The reply was: "Well, so has Mr. Bernstein, but perhaps if His Holiness would like to come down to a rehearsal, Mr. Bernstein could take a few minutes off to talk to him!"

Bernstein did finally have an audience with the Pope, and I assume that he (the Pope) lived up to expectation! But I can understand Bernstein's excitement at meeting a figure like the Pope; you see, he was somebody from a modest background, and he had "made it" and that was terribly important to him.

I remember after the Harvard Eliot Norton Lectures, which I must admit I didn't really enjoy—I didn't agree with the premises of some of the lectures—Harry Kraut said to me: "Would you like the printed edition of the Maestro's lectures?" I assented and he then produced a volume and said: "Would you like it signed by the Maestro?" I said: "Harry, please don't bother. I don't collect autographs." He said: "Oh, it's all right, I've got a machine that does it."

To return for a moment to Bernstein on the podium, how much do you think the visual aspect of a Bernstein performance—what Stravinsky and others referred to as the "chorybantics"—contributed to his success as a conductor?
It was fascinating. I heard him every week on Thursday night at Carnegie Hall, before they even moved to Lincoln Center. And if you went to a Bernstein concert—this actually happened the last time I heard him, at the Barbican, with the Concertgebouw—and you watched Bernstein conduct you were convinced you were listening to one of the great performances of all time.

But if you looked away, or shut your eyes, you realized that all sorts of things were being stretched, that all hell was breaking loose, that nothing was together. I had to edit his tapes, so I know about the inaccuracies.

It simply fulfills a theory I have always had about personal magnetism. Conducting has nothing to do with technique or knowledge; it's about communication. Stokowski once said that he could teach anybody to conduct in ten minutes, that you did it all with your hands and your eyes. Karajan had his eyes shut and scarcely moved. So did Böhm. Bernstein leapt about like something possessed. Toscanini was absolutely rigid from the waist down. Each conductor had his own style. It's a metaphysical thing.

If you ever watched Danny Kaye, in person, he had the same magnetism as Lenny. He would walk on and you loved him, you started laughing even though he hadn't said anything funny. If you actually take what he was doing it was minimal. But he had this personal magnetism. There are certain actors who can walk onto a stage, and they can be at the back of the stage, but you look at them. And there are certain conductors who have this kind of dynamic presence, and Lenny was the best example I can think of.

On the subject of Bernstein's recordings, where presence and choreography are not a factor, how do you think his interpretations will be viewed in years to come?

Well, I can remember putting one terrible question to the people in CBS during the seventies, when there was all the debate about whether we should let Lenny go. I said to them: "I have a little rule of thumb, and it's a very tough one. If you list any conductor, can you name one particular piece that you would rather have his peformance of than anyone else's?" I would take as an example the Schumann symphonies, and choose George Szell, almost without question. The Beethoven symphonies, being my age, I would choose Toscanini. Certain works of Mozart—Bruno Walter. And a huge repertoire—Karajan, whatever you might think of him.

I said to some of the CBS people: "Name me one work, apart from his own music, conducted by Bernstein, that you think you would rather have his performance of than anyone else's." I told them that the only one I could come up with was Sibelius's Fifth Symphony, which he did do superbly. It becomes a serious problem, when you can't automatically name something.

How about his performances of Haydn?
I would say Bernstein's approach to Haydn and to music generally was a little like Schnabel's Beethoven—he played it as though he was discovering it for the first time. There is no question in my mind that among the many marvelous gifts of Bernstein was the enormous joy, a visible, tangible joy in making music. So that when he conducted something like Haydn he knew enough about music to be classically a little restricted—with Mahler on the other hand, he might go mad—but he got inside the joy of Haydn and the incredible variety and imagination of his music.

I think he was totally uncynical when it came to music—I think he was just head over heels in love with it. For me personally, the day that Lenny died I remember feeling that I forgave him everything; he could be infuriating and irritating, and *musically* irritating, but at that moment I felt very grateful that he had been there, and that he had inspired so many people.

I think one of the reasons Lenny wanted to be recorded in concert was that at the moment of the happening it was always very exciting. Also, you see, I think John McClure didn't really help. The relationship between an artist and a producer—without wanting to bolster the role of the producer—is a subtle one. It depends on trust. Lenny wanted John, but he didn't always believe him, when, for example it came to a decision on a take. I remember attending some New York Philharmonic sessions in 1962 or 1963. In those days there was no closed-circuit television. They had finished a take and Lenny said: "How was that, John?" McClure came back and said: "Delicious." Bernstein turned to the orchestra, held his nose, pulled an imaginary chain and said: "Thank you, John, I think we'll just do another take."

Do you see Bernstein as one of the important figures in the democratization of the role of the conductor? His orchestras for the most part seemed to feel very affectionately toward him.

I think the conductor who did the most to democratize and—I hate to say it—to damage modern orchestras was Bernstein. The reason Toscanini and Szell and Bruno Walter and Karajan were so damned good was that they played an orchestra the way one plays a piano.

An orchestra is a body that has to be disciplined like an army. Why do you have square-bashing in an army? It's so that if you shout a command you can save someone's life. An orchestra has to obey a conductor's command. You know the definition of a camel; it's a horse designed by a committee. The democratization of the orchestra is to some extent what led to the destruction of great orchestras. You *need* someone in charge. Szell had a hobbyhorse he used to ride regularly with me, and that was that the seventies and eighties would see the dissolution of the great orchestras. He said that (a) there was too much democracy and (b) that conductors wanted to be superstars and wanted to tour around like virtuoso pianists and conduct a different orchestra every week. He would say: "Someone has got to stay home and do the five finger exercises."

Bernstein after 1969 was always a guest. From the moment he gave up the New York Philharmonic he was never permanently anywhere. And Harry, not unrealistically, said: "Look, I've got a superstar on my hands." Bernstein was, if you want, the first great virtuoso conductor who didn't want a permanent position.

But was that not because he wanted more time to compose?

Yes, I think he saw himself primarily as a composer, but as a frustrated composer because, somehow, the masterpiece never formed. I always had a mental scenario—just my fabrication—of how Lenny's life worked. I remember talking to Harry one day and he said: "Oh, Lenny has gone to Martha's Vineyard," more or less adding that he had gone in order to write a masterpiece. I always pictured Lenny spending the first day sharpening all the pencils, setting up the erasers, and setting up the scores, and looking out the window for inspiration. On the second day I imagined him calling all his friends to tell them that he had come to Martha's Vineyard to write his masterpiece. And the third day I pictured him saying: "Oh, the hell with this," and going back to the movies! That to me is what he was like. He was a brilliant but totally undisciplined man.

I think he just got out of the way of the self-discipline. I believe that you have to dedicate yourself—you can't be a sometime genius. I remember a program Bernstein did on Beethoven in which he talked about Beethoven and the battlefield of his scores, how everything was crossed out and rewritten and worked again and so on. Have you ever looked at a Bernstein score? Does it look like that? No. He had this wonderful, facile ability. He could sit overnight in a Boston hotel and write "I'm So Easily Assimilated" [from *Candide*]. But the kind of work, the honing, the polishing that made, for example, La Fontaine take ten years to write his fables—that I don't think really existed for Lenny. There were a few—the Rossinis and the Handels—who could knock out stuff at an incredible speed, but an awful lot of other people had to work and work at it.

I knew Sam Barber quite well. Sam wrote every day from nine until one and usually threw it all in the wastebasket. I once wrote to him when I was recording with Igor Kipnis to find out whether he had written anything for harpischord, and I said to him: "Would you mind looking in your wastebasket?"

Bernstein and Barber don't really seem to have liked each other.
No. You see, Bernstein was never part of the trilogy of Copland, Barber, and Menotti. It was again this thing of whether or not he was ever accepted. He had to sort of fight his way in and of course, Copland helped him a lot. But initially he was just the whiz kid pianist who accompanied Jennie Tourel, while the other three were already established.

Do you think that in some ways Bernstein outlived his time?
He did, I suppose. That was probably something else that Glenn had meant about the crisis of getting old. Bernstein at seventy looked silly doing what he was doing. And it was pitiful, the man [a Bernstein employee] by the side of the stage with the glass of whisky and the cigarette, for between bows. It's a story of self-indulgence.

Perhaps Bernstein was a parallel with Orson Welles, a parallel with a number of people you could name who were brilliant geniuses *manqué*, because they could only operate that way. I think he was a complex character who probably couldn't have existed without the excesses. From a musical point of view I felt that he was so *enthused* by the music that his own questionable taste sometimes got in the way. I wondered sometimes with Bernstein whether he knew what the real world was like. He lived in a world of his own creation. His immersion in the music was such that sometimes he almost didn't know what he was doing. It was a totally emotional experience for him. In a way, I suppose, it was like Glenn Gould singing—he couldn't help it.

NOTES

1. Murray Perahia has commented on the performances and recording of the Schumann as follows: "I enjoyed doing the performances of the

Schumann with Bernstein, because it was always alive, whether you agreed or disagreed with what he was doing. It was living music, and I really responded to that. But I had certain disagreements with him about tempi—for instance, it was a lot slower than I would normally have done it. I remember when we were trying to record it, the bassoonist came into the control room and said he couldn't play it at that speed—it was just too slow. And Bernstein said that he felt he was trying to do it faster! But he couldn't do it at any other tempo. I don't think it was something he could control in any way. I think he tried, but I don't think he could accommodate another personality or another concept that easily. I have had this experience with other great conductors, and as much as they try, their personalities are just too big, too strong. I felt a bit swamped."

Andrew Porter wrote a review in *The New Yorker* as follows: "The concert ended with an unhappy, incoherent account of Robert Schumann's Piano Concerto, in which Mr. Bernstein seemed to think he was the star and Murray Perahia, the soloist, a kind of continuo player. Since the conductor gets first go at most of the tunes, he can set the paces. They were set slow, and slowed still further by heavily sentimental touches of rubato. Mr. Perahia played along meekly and tidily. If he had taken up the themes at anything like a natural tempo, it would have sounded like a public rebuke to Mr. Bernstein."

2. Thomas Cothran.

IV
Bernstein Protégés

John Mauceri

John Mauceri, a Bernstein protégé, has a highly successful conducting career on both sides of the Atlantic. He has been equally at home in the opera house or on Broadway, whether conducting Wagner at Scottish Opera or the West Side Story Dances at the Hollywood Bowl. He conducted the European premiere of *Mass* at the Vienna Konzerthaus in 1973, which was recorded for television, and made his debut at La Scala in 1984, conducting Leonard Bernstein's opera, *A Quiet Place.* He was responsible for several structural changes to *A Quiet Place,* with the acquiescence of the composer, following its world premiere in Houston in 1983. He has also played a crucial role in restoring much of the original music to *Candide,* and has performed this and other Bernstein compositions extensively in Europe, Israel, and the USA.

Bernstein's critics have always held that he "spread himself too thinly," that he should have settled for one particular speciality. Ned Rorem has remarked that this would have crippled Bernstein, both musically and personally . . .

Yes. I remember reading an interview with Lenny in which he said that whenever he entered or exited a country he would fill in on his passport form not composer or conductor, but *musician.* Of course people in the press spent a lot of Lenny's life telling him what he should have done: he should have been a concert pianist; he should have composed more; if only he had stayed with the Philharmonic longer. And people wouldn't let him live his own life. But he created his own career, in his own image.

Since Lenny was so overwhelming—threatening to some people—he spent a lot of his life getting to read unfavorable stuff in the press. And it would be naive for anybody to think that there are any artists who are not hurt by negative reviews and unfair publicity. The critics who destroyed Maria Callas, who destroyed George Gershwin, have a lot to answer for. All of us fight it, and pretend that it's not there, and get up and say: that's the way it is. But something does get taken out of you, something does begin to bleed internally, something starts to happen. If you look at artists and the kinds of things they have to put up with, the stresses that they have to deal with are enormous, both publicly, whether performing music or writing it, and also personally, because that is a very lonely journey.

Bernstein seems to have had a long-suffering attitude toward the New York critics. Did he ever speak to you about this?
He talked to me about [Howard] Taubman and a little bit about Schonberg. He would say: "Imagine what it was like to wake up on Friday morning after having conducted your first concert each week [on Thursday] and while you're having your breakfast to read all of that stuff and then to go to the concert hall and conduct the matinee." I think it hurt him terribly.

In the last decade of his life, though, there seemed to be a critical change of heart.
Yes, in the last ten years he only got great reviews. Something happened. His hair went white! Something changed, and one thing that we can say was that in the last ten years there was almost no demurral from anyone. Whatever he did, whatever exaggerated tempos, or experimentation, he was basically permitted by his public and the critical community to do whatever he wanted. And from that point of view, I suppose, we can all be happy for him, that the fight was over, that in one sense he outlived the older critics who set him up and then tried to destroy him. I don't think he was necessarily better in the last

ten years than he was in the rest of his life as a performer. I just think that some kind of international *détente* was reached!

Bernstein's tempi as he grew older became markedly slower, sometimes exaggeratedly so. Would you say that this slowing down applies to a majority of older conductors?

No. I don't know whether generalizations are possible here. Toscanini probably got faster. My first conducting teacher, Gustav Meier, said that the ones who start out fast end up slow and vice versa. Lenny did become slower, but he also admitted to me that, for example, the Brahms Third [with the Vienna Philharmonic] was too slow, that it was an experiment. He conducted that in six as opposed to in two, and that was a "Lenny six" with all the subsidiary beats. He felt in retrospect that it was too slow, but that it was just a view of it. I think Lenny understood that a performance is a *view* of a piece of music; it *isn't the piece of music*. And most people get confused with that. The confusion is that "the work is equivalent to my favorite recording of the work."

I think there is something here with Lenny that is rarely discussed and that is that Lenny in a recording studio and Lenny in a concert hall were two very different people. Very different in the sense of how to use the room and also the medium. Lenny in the studio tended to be slower, because, like all of us, he wanted to hear everything. It also depended, obviously, on the acoustics of the room and the microphone placement; if the room was dry he tended to conduct faster; if the room was reverberant he tended to conduct slower. This is why the last recordings, which are mostly based on live performance, give one a better idea or image of how Lenny performed in public.

Many people have commented on the memorable qualities of a Bernstein concert. How important do you think it was to hear him in the flesh?

I think you can never understand any artist or musician unless you hear him or her live. And this was particularly true of Lenny. Recordings simply document a day that was had in the studio—or five days. Erich Leinsdorf once said, and brilliantly so, that the most powerful tool a conductor has is silence. And in a recording that is meaningless. A performance is an agreement between an audience and the performers. It has to be. You cannot have a great performance without the acquiescence of the audience.

How about the Unitel videos? Do you think they capture any of the crackle of a Bernstein performance?

I don't think the particular aura, or humanity, of Lenny that was experienced in a live performance can ever be captured either by recordings or by television or videotape. Television is the medium that comes directly into our homes, but is by the same token the most removed from us and the most detached. There's a three-inch speaker and a twenty-inch screen, and you can watch how Lenny perspires, or how he smiles, or how he ages, but you're detached. There was a *danger* involved in a live performance of Lenny's. Lenny, for example, might conduct Mahler Nine and it would be ten minutes longer than you had ever heard it before, due to the extraordinarily slow tempos he chose to use. These performances were *dangerous*, they risked chaos, and don't forget that. There was no necessarily implied happy ending.

To move on now to the compositions, you have probably performed more of Leonard Bernstein's music than any other living conductor and have acted as a catalyst in the completion and/or alteration of works such as Mass, A Quiet Place *and* Candide. *Could we begin with talking about* Mass?

I think the creative process of *Mass* was very complicated. Lenny had a sort of writer's block during the period of its composition. It's a huge work and a difficult and painful experience putting it on. You can say: "Sing God, a Simple Song" [the words of the opening number of *Mass,* sung by the celebrant], with two church organs, a children's choir, a band, a string section, nine percussionists—some simple song! Of course, unlike many of his other theatre works—*West Side Story,* where Jerome Robbins had equal and in some cases more powerful authority than Lenny, or *Candide* where I think there were more collaborators than ever before in the history of music theatre—*Mass* was Lenny's creation and as such he had authority over it, even though he had been helped with some of the words by Stephen Schwartz.

Mass *has been described as a "melange of styles" (Harold Schonberg). Do you think it holds together, structurally?*

Well, *Mass* is in a sense an accumulation of fragments; for example, "Simple Song" was originally conceived for a film about St. Francis by Franco Zeffirelli, which became *Brother Sun, Sister Moon.* Lenny never went ahead with that project, because the more he and Franco talked about it, the more it seemed they were talking about two different works. Then there is "Sanctus, Sanctus, Sanctus," which Lenny wrote for Helen Coates [Bernstein's secretary] for her birthday; it is really "Happy Birthday, Helen Coates." All of us who knew Lenny well would either get a poem or a song for our birthdays, and sometimes these would reappear in larger pieces.

Mass began my work on editing his music, and what was a surprise to me was the bad state his music was in, in the sense of dynamics, articulations, and the state of the parts—just getting an instrumentation list for *Mass* seemed to be impossible; were there five trumpets or four, for example?

I am fully prepared that there will be those who tomorrow, or in thirty years' time, may feel that my influence on Lenny perhaps confused his artistic vision. I would say to them that people had to help Lenny to write. He always had an assistant working with him, and he needed one. I would say to Lenny:

"This is silly; Let's go through this from the first bar to the last bar, and correct it and make it work." I would bring fifty or so questions to Lenny, and we would sit in Fairfield, Connecticut [at Bernstein's country home], and we would do it.

If you look at the case of the *West Side Story Dances,* they will be ready, fully corrected, in the summer [1992]. When I conducted the *West Side Story Dances* in Los Angeles in the late seventies, I borrowed Lenny's score, and there were little red x's in the corner of it, something like one hundred and four of them, one hundred and four corrections! And not just a *forte-piano* instead of a *forte,* but notes, a G flat instead of an A flat. I showed up in Los Angeles and none of those corrections were in the parts. And there was no way on one rehearsal at the Bowl that one could begin to do anything except to conduct it. But, worse, it meant that performances of the *West Side Story Dances* all over the world had at least one hundred and four mistakes, and that's to start out with, not to mention those that might accidentally be added in performance! So this kind of thing happened a lot. I would make a phone call to New York and all hell would break loose. It seems to me that part of my role with Lenny as an assistant and a colleague was to help put the music in order.

Could you talk about how Bernstein marked his scores?

I often borrowed his scores; for example, when I conducted my first Sibelius Fourth Symphony I would borrow his score and talk to him about it, and I learned to mark my scores in a very similar way. He would use a combination of blue pencils and red pencils, blue for the cueing of instruments, red for any editorial changes. This way of cueing instruments was an accumulation of things he had learned from Reiner, and also from Mahler scores. So, I learned a lot from him, needless to say. Toward the end I felt I was returning the compliment. I remember before he did *Bohème* I gave him my personal score, and we sat for four hours and went through it from beginning to end, and it was a wonderful thing to be able to do. In a way the closures, the various closings, began about five or six years ago. I see that in retrospect. I would come up with a new long joke for *him* to laugh at, instead of vice versa, and also I brought him some Hindemith symphonies which he had never heard before, and Shostakovitch's last symphony, No. 15, which he thought he would never conduct, but later decided that he would.

Could we move on to A Quiet Place, *and your role in the changes that followed the world premiere in Houston?*
Lenny began *A Quiet Place* in the wake of Felicia's death, and Stephen Wadsworth [the librettist] was also working on the opera in the aftermath of a death—his sister's. They worked for about a year at Fairfield and then friends of Lenny would get to see bits of *A Quiet Place* rehearsed and performed with piano and singers in the Dakota [Bernstein's New York apartment on the Upper West Side] and it was often overwhelming. It was one of the most complete, large-scale works that Lenny had written.

In June 1983, I flew with Roger Stevens[1] to hear the world premiere of *A Quiet Place* in Houston. The evening began with *Trouble in Tahiti* in a fairly camp and very funny production, and then came *A Quiet Place* and the audience *hated* it. It was alienating and unmoving and a deeply depressing experience. I was very troubled by it, because it had been so good in the house. This happens a lot in music theatre works, where at rehearsals or final run-throughs, with the whole cast and piano, people are weeping and cheering, and then you add the orchestra and the lights and the costumes and there is a terrible moment when the show sinks. *1600 Pennsylvania Avenue* was one of them, and *A Quiet Place* was another.

I thought about it a lot, discussed it with Roger, got timings and so on. It really hurt me because I knew that Lenny had composed one of his most complete works, a long and fully thought-out piece, which represented the various complex emotional states of mind of its composer. The piece *was* Lenny; his life, or representations of his life were up there on stage. Every other line is a direct quotation of something that happened to him or to his family, something that maybe Alex said to Lenny, or Nina said to Felicia—it's all in there. So in Houston, the people were rejecting not only his work but also his life.

After a lot of thought about the piece I came up with the idea of the flashback. What became clear to me was that the harmonic language of *Trouble in Tahiti* was so accessible that the first scene of Part Two in *A Quiet Place* with its complexity of musical language was something that the human mind could simply not adjust to. It was like suddenly being in Strauss's *Elektra* after having heard *La Gazza Ladra.* It just didn't work stylistically. So I told Roger about the flashback idea and he liked it and then I phoned Shirley Bernstein [Bernstein's sister] and I remember I woke her up. She said to me: "I may be asleep, but I recognize a great idea when I hear one." And I spoke to Stephen [Wadsworth] and he liked it, so then we took it to Lenny, who of course fought it. I said to Lenny: "It's not working like this." And he knew that. So ultimately he agreed to make the changes and rewrite some of the music.

Following these changes, you conducted the European premiere of A Quiet Place *at La Scala, Milan. Could you describe this?*

Conducting *A Quiet Place* was everything I knew it would be. There were not only a lot of wrong notes in the parts, but the piece seemed to have been orchestrated for Godzilla and King Kong to sing. Lenny had only supervised a lot of the orchestration with the help of Sid Ramin and Irwin Kostal. In Jones Hall, in Houston, they fixed that with the microphone levels, but here we were at La Scala. You had four horns going *fortissimo* and the percussion going beserk, and some person trying to sing. So I had to reorchestrate the piece in the sense of cutting out instruments and making it work on the stage. I would ask Lenny's permission about things, but he was not really in a position to be helpful. He was sitting there with the score, about twenty rows back, in an emotional state. Then I had to try to explain, in Italian, to the orchestra, what *A Quiet Place* was all about and I remember the tuba player saying to me: "Aren't there any *normal* people in this opera?" And I said to him: "Only you!"

And so we all did the best we could to make the work stageworthy. Stephen [Wadsworth] directed and in the end we got excellent reviews and the performance was a big success. Lenny went out of his way to give me credit for the work we did. He said: "I could never have done this without these two geniuses, Stephen Wadsworth and John Mauceri." But there were two sides to that; if it had been a failure he would have been protected, he would not have been implicated.

Do you think that Bernstein would have written other operas after A Quiet Place *had the initial response been more favorable?*
Absolutely. The response was negative, and swift and clinical, lethally so. When Lenny was writing *A Quiet Place* he was so happy he thought he would only write operas for the rest of his life. And of course he never wrote another one. Now that he is dead everybody writes about what a great man he was. It's the most bizarre relationship that critics have with creative artists and performers. And it will never change. One doesn't understand what is at the heart of it. What is at the heart of all those years of Harold Schonberg waiting until Lenny resigned from the Philharmonic to write a review saying: "Too bad he has resigned—he's only just now getting good." [In a Sunday essay entitled: "BERNSTEIN: WRONG TIME TO LEAVE?" Schonberg had noted that Bernstein's conducting "now seems more intent on substance and less on flashiness . . . . In short he is threatening to turn into the kind of conductor that his

talent originally indicated. Therefore—a typically Bernstein gesture—he is leaving."]

If Lenny were alive right now and writing another opera, or if we were sitting here talking the day after his new opera had just premiered, we would be reading some pretty nasty stuff. The pattern repeats itself. You should see what they were writing about George Gershwin in 1936. All the serious music critics were saying that he had no right writing operas and concertos, and the commercial guys were saying that he had lost his commercial touch. By the time he died he had no constituency.

Could we discuss Candide? *Your role in restoring much of the original music to* Candide *seems to have been a crucial one.*

First of all I think it's naive of some people to feel that the 1956 version [the first] of *Candide* was the best and was simply misunderstood. All you have to do is read Lillian Hellman's book and imagine Lenny's bright and witty music, full of references to European operetta, next to a very heavy, unfunny, and didactic book. It was impossible. Lenny was clearly writing a different piece from Hellman. Some of the best music was never even heard in '56 or in '57 in London. A little of that music at least got into the Chelsea Theatre version which I did with Hal Prince in '73—"Candide's Lament," for example, and the "Barcarolle" which underscored the Old Lady's Story. It's interesting, Lenny often seems to have refused to fight for his own music. I can't explain why that is, maybe he was just unsure. In *West Side Story,* for example, Jerome Robbins always had the stronger hand.

With *Candide,* in the '73 version, Hugh Wheeler wrote a new book which was closer to the Voltaire and I was given the so-called "trunk" music, which was all the music that Lenny had ever written for *Candide.* From the point of view of versions this was probably more difficult than *Don Carlos.* There were three versions of the "Auto-da-fé," with much of the same music, there were two Syphilis Songs, and so on. I had to decide where some of this music would go, and so in a sense I became the composer without having written the music. It was like trying to put together a strange Chinese puzzle. And it was limited by the fact that the show was supposed to run for less than two hours as a one-act musical. Then the song "Eldorado" could not be used at all, because Lillian Hellman wrote the lyric. The legal agreement with Lillian was that she was bought out of *Candide.* Not one of her words could be used, or the locales.

There followed, after the Chelsea Theater and Broadway version, an Opera House version and finally the version we put together at Scottish Opera, with the help of John Wells, where nearly all of the original music was restored and we got permission to go to the right places i.e. all the French music could take place in Paris, all the Italian music—the "Barcarolle" and the Venice Gavotte—could take place in Venice, and so on. We had Jonathan Miller as director and he brought a European sensibility to the whole thing, which it needed—it is, after all, Lenny's most European score. In the American productions the Pangloss character, for example, had been a little too much like Groucho Marx.

Lenny came to the Scottish Opera performance and loved the new version, and that, with one or two small changes that he made, was what he used for the concert performances and the recording in London [1989]. I think that he died with the feeling that *Candide's* time had come, that it was an important work.

In spite of Candide, *do you think that as Bernstein grew older he felt an increasing sense of frustration as a composer?*

I think a lot of his last works are a series of short pieces, because he couldn't apply himself to writing longer works. I'm thinking of the *Divertimento* and the later songs, for example. I think his unhappiness had to do with his inability to compose *more*—

that was the great frustration. That is what he hoped to do even when he retired from conducting, that he would at least compose. I think the reason that he didn't write more was that he didn't really like being alone. Composing is such a solitary process. For Lenny to go from the public sharing of ideas, the public forum, to the private forum in which he had to deal just with himself, and choose notes at two or three in the morning was a well-documented and very painful transition. He would take half a year off from conducting in order to compose, and often get nothing out of it and then become severely depressed.

I think also Lenny was terrified that his music might be boring—I think he was more afraid of that than anything else. Often in the seventies, when I was conducting his music, he would tell me that it should be faster, that it should move right along. And yet when he conducted it himself he would sometimes milk it for all it was worth. I think at that time he had doubts about whether what he was writing was any good, but that seemed to change in his last years.

How easy do think it was for Bernstein, once he had taken time off to compose, to move back to a period of conducting?

It was difficult, I think. I remember, he would often feel dreadful before going off to Vienna. He would feel that he didn't know his music, or he would be worried about getting himself up for the public performance. And then when he was in the middle of doing it, he seemed quite happy. And every experience seemed to be the best. I always found it touching that he would come back and tell you how long the ovation was, or how big. Or he would say that he wished you had been there when he was rehearsing a particular work. He would be full of enthusiasm.

You know, I think at the end of the day what he was most astonished at was his own celebrity. Don't forget he was a small-town boy, born in Lawrence, Massachusetts. I remember how *enormously* excited he was at the prospect of meeting the Pope.

What could one say? Or he would tell me a story of how Michael Jackson had come to one of his concerts. I suppose a lot of us are basically small-town kids who grow up and are astonished to find that the people that we are working with are famous.

Do you think Bernstein was more socially conscious than other musicians? For example, he often concerned himself with events outside the concert hall, usually of a political nature.

His rage was mostly directed toward world events, toward the cruelty that he perceived to be happening around him. His philosophy (and his music) was about allowing people to be themselves, to get on with their lives, to be a little different if they wished to. I would say he was angry with very few musicians—Herbert von Karajan principally. Lenny had brought Karajan to America and he was angry that Karajan's behavior toward him had not been more collegial. Apart from the fact that Karajan had been a Nazi, I remember Lenny saying that he didn't think Karajan ever read a book, that he was a complete non-intellectual, and that bothered Lenny terribly. But Lenny was usually very generous toward other musicians.

Turning to Vienna for a moment, how do you explain Bernstein's long love affair with that city?

It is one of the many inconsistencies in Lenny. Why Vienna, and why Berlin? I will say this, that his credibility in America quadrupled when Vienna accepted him. In a sense that was a turning point; it was the same Lenny who had written *West Side Story* and conducted the New York Philharmonic, but when he was accepted in Vienna there was a new sense of recognition.

In America his career was probably helped by the advent of television and the long-playing record. Everything had to be re-recorded when the LP came along. That and the fact that television existed for the first time, that it was after World War

II and there was a strong American feeling, all of those things helped make him bigger than he might have been twenty years before or twenty years after. It doesn't matter though—he was an extraordinary musician.

Also, of course, Lenny loved to teach. In fact he frequently learned by teaching. Lenny's capacity for sharing was one of the things that drew me to him. He loved life and he loved people. He was always interested in learning and then sharing what he had learned. Whenever either of us discovered something new we would immediately be on the telephone to tell the other. My relationship with Lenny was very much based on story-telling, on jokes, on laughter. Lenny and I collected what I called life-jokes. They were sometimes funny, but they were all about life. I made a short speech at his Seventieth Birthday Concert and I said that what I would I think of most about him was the telling of stories.

In terms of the repertoire Bernstein covered, do you not think there are some surpising omissions? For example, there is no evidence of his ever having conducted a Mozart opera.

I think near the end of his life he tended to conduct the same pieces. Which is fair enough. But we all wanted him to conduct a Mozart opera—imagine, he never conducted *Don Giovanni* or *The Magic Flute* and yet he conducted all those Haydn symphonies. His interest in Mozart seems to have been nil. He had also very little interest in Bruckner. Mahler was his man. Then there was Beethoven, some Brahms, and occasional American works. It's quite an interesting legacy. Also I think he will live on in the minds of the thousands of musicians who played under him, who worked with him.

For me, personally, he is with me everytime I put a blue pencil into a pencil sharpener, everytime I study a new score (or read a new book) and, most of all, if I ever drink American Scotch—the smell of that and the tinkling of the ice can only mean it's five o'clock and I'm working on a new project with Lenny.

NOTES

1. Roger Stevens had been the principal savior of *West Side Story* in the fifties—he had raised money for the project when no one else had wanted to produce it. In addition to other Bernstein projects, Roger Stevens commissioned *Mass* and was a co-producer of *A Quiet Place* for the Kennedy Center.

Justin Brown

Justin Brown, a Bernstein protégé, had his first major professional break when he conducted *Mass* as part of a Bernstein Festival at the Barbican in 1986. Subsequently he studied with Bernstein in Germany and won a scholarship to Tanglewood, where he also took part in masterclasses with Seiji Ozawa. He worked for a time as assistant conductor to John Mauceri at Scottish Opera, and guest-conducting engagements have included work with the London Symphony Orchestra (in a critically acclaimed performance of *West Side Story*), the London Philharmonic Orchestra, English National Opera, and the Santa Fe Opera. He was Leonard Bernstein's assistant for the performances and recording of *Candide* in London in December 1989.

Could you describe your first meeting with Bernstein?
In June 1985 I think it was, Lenny was doing Mahler's Ninth with the Concertgebouw at the Barbican, and he came to a performance of John Mauceri's at English National Opera, where I was working as a répétiteur. That was where I met him for the first time. I was wild about the Mahler and he and I hit if off immediately—we discussed the work for hours at the Savoy, where he was staying. Lenny was happiest in many ways when working with young people; first and foremost he was a teacher. Harry Kraut, Lenny's manager, then asked me to come to Vienna, that October.

Every year Lenny went for a few weeks to Vienna, where he would do maybe two or three concert programs which Deutsche

Grammophon would record and Humphrey Burton would video for Unitel. So I went for two weeks. He did an all-Schumann program and a very bizarre program of the Beethoven Choral Fantasy, a Haydn symphony, and the Shostakovich Ninth. Lenny was thinking about making a biographical film on Schumann's life at the time and I did some research into this.[1]

He was often at his most alert between midnight and four a.m. He always had a piano in his hotel room, always had one shipped in. By the time I met him it was a Bosendorfer; before that it had been a Baldwin I think. We would play music—I remember one night we played all fifty-seven Chopin mazurkas and I went home at about seven a.m. We would talk for hours.

In 1986 you conducted Mass *as part of the Bernstein Festival at the Barbican. How did you feel about the work?*

Well, conducting *Mass* was my first major break, my first professional conducting venture. When it came up as an idea I thought it was just impossible. I couldn't see how a bunch of mid-eighties Guildhall music students would ever accept the work; London in the mid-eighties was a very cynical place, and New York in the early seventies was the exact reverse. *Mass* is the most eclectic, outrageous, flower-powerist piece you can imagine. I couldn't see how it would work, and I was very worried about it. I didn't like it as a complete work. It's such a mess; I suppose that's the unfair way of putting it. It's a mish-mash. Most of the music is immediately gripping, but put it all together and it's very difficult to see how you can bring it off.

I worked very hard at it and as usually happens when you work on anything to do with Lenny, you get drawn in. Lenny came to about four rehearsals and two of the performances. He was very fond of the piece. I don't think he would have been able to look you in the eye and say that it was his greatest work, but it was the one that nobody could understand and he felt protective toward it. I think he felt protective toward the pieces of music that people didn't want to play. It was very fatherly the way he would look after the weaklings in his family; he often spoke about this, and would talk at great length about the pieces that people didn't like.

Do you think Bernstein resented his Broadway image, and the fact that his serious compositions were never fully accepted?

There was that feeling, but I can't say that it surfaced that much; he was certainly not bitter, he was the least bitter of people. He might say, if he had read something bad in the press: "Those assholes, they just don't understand." Sometimes he would become completely outrageous and say, for example, that *Divertimento* was the greatest work he had ever written. That felt as though it was because he wanted somebody to perform it. *Divertimento* is clever, well written—it took him longer as he grew older to produce the stuff, and took a lot out of him—but it certainly isn't his greatest work. That aside, I never got the feeling that he resented the fact that he was famous for *West Side Story* in the way that, for example, Constant Lambert resented being known for *Rio Grande* or Gershwin for *Rhapsody in Blue*. I think he knew perfectly well that *West Side Story* wasn't just his most popular piece; he knew that it was a hell of a lot more than that, that it was a unique piece of music theatre, and completely deserving of its popularity. He didn't talk about *West Side Story* much. Maybe it's because it has been talked about a good deal that he didn't seem to need to.

To return to Mass *for a moment, it seems to work best primarily as a theatrical experience.*

I think the audience [at the Barbican performances] got drawn in partly because the entire building was full of performers. The way it was staged was that I was in the middle of the room, and the orchestra and chorus and performers were everywhere around me. To conduct at 360° is very rare and so from that point of view it was an extraordinary experience. The piece finishes with a prayer, which is sung by the whole chorus (150) and the street chorus (30 or 40) and the entire string section. I gave one downbeat and they played and sang the entire chorale unconducted, just by listening; there was no light by this point, just a few candles and everything became completely black. It was an extremely intense moment.

You won a scholarship to Tanglewood shortly after Mass. *Could you describe your work with Bernstein there?*

It was fascinating, but you learned about music, not about conducting. At Tanglewood, Lenny was sort of an event. He was there for two weeks and he took the place over. You didn't feel that he was a member of the conducting faculty as such or that it was really organized. Ozawa was there too. He is technically very strong, but doesn't speak English all that well, and he didn't do very much teaching. Nevertheless, just by watching him you can learn a lot. That's often true with conducting; you learn as much by watching as by being told where to put your hands. Lenny always said it was just not worth talking about that. I think the most important thing that Lenny taught was the need to give everything of yourself to the music. He was not interested in conducting technique as such. He said that he had never really worked out what he did—he just did it. This is not strictly speaking true, because of course he studied with Reiner. But he saw his function at Tanglewood, I think quite rightly, as being to talk about music. There were other people telling you how many beats to beat in the bar, where it was a good idea to subdivide, and why you shouldn't do a particular passage at ♩ = 88. Lenny felt that he was there to cut through all that; he was always telling people: "Don't conduct beats, conduct music." At the time it confused me. At that age and at that stage it was very difficult to make that work, because of course you do need technique when you're conducting. You've got to work out your own technique and feel at home with it. Then you can let it go and do what he is saying.

Could we look at Bernstein's own technique for a moment? It seems to have been a very personal technique.

It's very hard to talk about his technique; it seemed to depend so much on what was happening in his head. Sometimes even though you knew that he was thinking and feeling and expecting this great music to come out of the orchestra, you would look at his hands and wonder how people played at all. But they did. All they seemed to need was a hint from him, of how he wanted things to go, and they would be there with him. They trusted his ideas and he rehearsed brilliantly.

It didn't always work; there were performances which didn't quite come off, places which were ragged sometimes, like anybody else. The power of it was not in the technique; rather in the use of his brain. Of course, he used his whole body, he would jump around, but it wasn't something you would want to emulate. Kleiber, for example, uses his whole body too, but he uses it in an extremely well-organized way. He's a model of balance, every gesture is beautiful and every gesture perfectly reflects the music. That wasn't true of Lenny; every gesture of Lenny's reflected the way he was feeling. I don't think it was superfluous, it was just that he was feeling a lot. I also don't think it was for the audience. He was an enormous showman, of course, but he was a performer. He didn't feel that there need or could be a distinction between an artist and a performer. I think Kleiber or Karajan were less interested in the audience, less aware of the audience than Lenny, but I don't think that is the same as saying that Lenny did histrionics for the benefit of the audience.

Bernstein often spoke about looking at a score from the point of view of a composer rather than a conductor. He sometimes referred to interpretation as a form of "re-composing."

I think people maybe misunderstand what he meant by this. It is absolutely true that he looked at the score from the point of view of the composer. He understood the process of composing because he was a composer and in that sense he had an insight into the composer's mind which perhaps non-composers may not have. Every minute detail of the analysis of the composition he would go into. As far as "re-composing" is concerned sometimes he might re-orchestrate a passage slightly, or bring things out that other people didn't, or he would disobey a composer's tempo markings or other indications, but that was as far as it went.

Many conductors have done rather personal things; the Czech conductors often do things that are not in Dvorák's scores, because they know that that's the style and that's how it should be. A lot of the force of Lenny's interpretation seemed to come from his ability to get inside the composer's head. Maybe that's because he was a composer or just because he was a great musician. I get that feeling from Oistrakh or Rostropovich—you get the feeling that they are somewhere in there. And Horowitz, definitely—he was very similar to Lenny in that way—he would disobey a tempo marking completely, but in the most logical way.

Bernstein always had a strong identification with Mahler—both with the man and the music—but his attempt to draw parallels between himself and Mahler seems to have resulted eventually in a kind of self-delusion.

Lenny of course loved and understood Mahler's music, and he conducted the first complete Mahler cycle on record. I think he saw a great affinity and a cause to champion, but it didn't ever seem to me as though he was trying to say he was a reincarnation of Mahler. They were very different types of people: Mahler was this wiry, ambitious, paranoid person; he had qualities which couldn't be more different from Lenny. If Lenny was ambitious it was subsumed under the weight of his enthusiasm.

You never got the feeling that Lenny was political in terms of the business side of the music world, whereas Mahler was extremely political. It would be naive to think that Lenny was not *au fait* with that whole side of things, but by the time I met him he was employing Harry [Kraut] to do all of that and Harry was brilliant at it. By that time he simply didn't need to use the politics of the music business for his own ends. He was who he was; if one had known him in 1950 the story might have been completely different. He always said to me: "I never was an ambitious young conductor, besides which I never wanted to be a conductor; I just wanted to be involved with music somehow." I think that is slightly rose-tinted, but he often said it.

I think throughout his life what came across was a massive amount of energy and an extremely focused brain. I have never encountered anybody with a memory like his, and the reason for that great memory was his concentration. For example, with people, if he met somebody new he would give them one hundred percent of his concentration. He would remember everybody, people from all around the world, orchestra players or just people he had met. People felt that they occupied part of his mind and his affection. For example, I knew that there were any number of young conductors who knew him and were friends with him and benefited from him, but that didn't matter. You rang him up or you were with him and you were absolutely in the forefront of his mind, and he was totally involved in his friendship with you at that time.

Of course his musical memory was also unbelievable. He and Adolph Green when they got together were hilarious. They could remember the lyrics and tunes of every popular song written in America, from 1910 until today. They would play games; one of them would sing (with no words) the bridge or the verse of a song, and the other would have to guess the song. They were quite phenomenal at it.

To move on, Bernstein was adored by the Viennese public. In what is reputed to be one of the most anti-Semitic and conservative cities in Europe this seems strange.

Yes. I don't understand it at all. The Viennese have always had a very strange relationship with their Jews. At the turn of the century there were many great Jewish intellectuals in Vienna—Mahler, Schoenberg, Berg, and the painters, and Freud. Vienna initially turned its back on them and then embraced their work after they were dead.

Mahler was for a time popular as a conductor, I think. When Lenny first went there it didn't work out and it took him a long time to come back. When he did he was a hero. But the Viennese are very, very right-wing, very stuffy, very gossipy. Very unlike him. Lenny came in there and more or less took the place over. The idea that they would trust somebody like him to tell them about *their* composers—to perform Beethoven and Brahms and Mahler—is extraordinary. I don't know the Viennese well enough to know why there was this particular relationship with Lenny. When I was there with him we met countless people who absolutely adored him. There would be many parties for him and so on.

Do you think he felt comfortable in Vienna?

Well, if you're lionized I think you tend to feel comfortable. Mind you, the year after I was there he gave up on the Sacher Hotel. But whether that was to do with their "Vienneseness" I don't know.[2]

Could we talk about "placing" Bernstein as a conductor? With whom does he belong in this century?

There's nobody really. Mitropoulos perhaps. And Koussevitzky. He would like to have thought Koussevitzky. I was once with him in Tanglewood when he was listening to a performance on the radio of Sibelius Second and he said: "This is me isn't it, it must be my old recording. This is exactly right; it must be me!" We listened to it right through to the end and it was Koussevitzky. Lenny was actually doing the work that year at Tanglewood. I think he was quite pleased, secretly, that it was Koussevitzky. But no, there is nobody in my knowledge who was anything like him; I think that's true of any real original. Karajan, for example, seemed to embody the great German tradition, but you couldn't really trace the individual influences.

If one compares Bernstein and Karajan, the former seems always to have a feeling of life, which at times goes over the top, whereas the latter delivers highly polished, very beautiful, and sometimes almost mechanical performances.
I think that's a bit of a generalization as far as Karajan is concerned. His performance of the *Liebestod* with Jessye [Norman] is one of the most extraordinary pieces of music-making I have ever seen. It's full of humanity. On the other hand if you watch say the *Pathétique* film which was released in the mid-seventies with Karajan, it is mechanical—he *was* capable of that. Lenny was always full of humanity, sometimes, as you say, more than the music would want; Sibelius, for example, never really sounded quite bleak enough.

Do you feel that as Bernstein got older his tempi became indulgently slow?
Yes, they really did. By the end he was quite phenomenally slow and I don't know whether that was to do with the speed of his mind or what. Sometimes you got the feeling that he wanted to milk every single note because he loved it so much, and that could become difficult to take. Karajan seemed to conduct faster; his last set of Beethoven symphonies is often pretty brisk. But often with older conductors this slowing process does seem to happen and it was certainly the case with Lenny. I think it's also true with Lenny that you had to be *there*; there was something he generated in the atmosphere of a concert that was completely memorable and that is not always recaptured on CDs.

Bernstein has probably been the most famous conductor of his generation and perhaps of the century. How do you explain this magnetism that he had for the public?
I think because people want optimism. I think the more cynical the age becomes the more extraordinary you become if you aren't cynical. Lenny had this flood of unfashionable brotherhood and openness and optimism. He wasn't always optimistic; he got depressed. But when he got depressed it was usually for the state of the world—it was mega-depression. I think people loved him because he seemed to spread an air of something worth living for. And he was an unashamed populist as well—he was always giving out to everybody. For Lenny there was no point loving the music unless he could share it, and that's why he was a great teacher. I got the feeling that he would really rather have had somebody in the room looking over his shoulder while he was studying his scores, that he would have more musical ideas if there was somebody to tell them to. He wouldn't want to be in a room on his own.

In some quarters it has been said that Bernstein's last years were characterized by despair. Now that goes against almost all you have been saying.
I didn't really see evidence of despair. I think during the illness things got very, very difficult for him, because he was miserable not to be able to live to one hundred percent. Prior to that, in Vienna, I remember he was on a low because of what was happening in the world, politically and so on. Then after that I saw him happy again; he was happy when he was making music and when he was teaching.

You were Bernstein's assistant on one of his last major projects, the performances and recording of Candide *in London in 1989.* Candide *has had a somewhat checkered career. Did he ever discuss this with you?*
Lenny sometimes talked about the rows he had had with some of the librettists [these included Lillian Hellmann, Hugh Wheeler, and even, for some revised lyrics, Steven Sondheim] or

a particular number that they disagreed about. I think he found Hellmann rather an impossible person, but so did everybody else. I went over to New York and we rewrote several parts of it and cobbled it all together and made it work. He got quite into rewriting it and changing and discovering things. He was an immensely practical musician.

He was very ill during the recording; he never really got better after that. Several of the cast caught flu and some people were unable to make the recording sessions; Christa Ludwig didn't make any of the sessions and we tracked her in later, but maybe I'm not supposed to admit that. Lenny really wasn't well. They made a film of the concerts with a great narrative by him in which he elucidates the whole business of Hellmann and the history of the piece. He's obviously not well in that. It was sad, but he still managed to pull out a lot of vitality, despite it.

NOTES

1. Bernstein did not go ahead with this project, as there seemed insufficient evidence to support its subject—Schumann's alleged homosexuality. When Bernstein made films about Mahler and Wagner (the latter incomplete), he would invariably attempt to elucidate in the lives of these composers aspects of himself.

2. Bernstein may have left the Sacher Hotel because they redecorated his suite without consulting him.

V
From the Theatre

Jonathan Miller

Jonathan Miller, the English director, began his theatrical career on the stage in the 1960 revue *Beyond the Fringe* at the Edinburgh Festival. He directed his first play, John Osborne's *Under Plain Cover* for the Royal Court in 1962. In 1968 he directed Sheridan's *The School for Scandal* and since then has concentrated on Shakespeare, directing *King Lear* in Nottingham in 1970, and in the same year *The Merchant of Venice* with Olivier as Shylock, for the National Theatre at the Old Vic, and *The Tempest* at the Mermaid. He was Associate Director of the National Theatre from 1973–75 and Artistic Director of the Old Vic from 1988–90. Miller has also been prominent as a director of opera, with many fruitful productions for English National Opera, and direction that is notable for its striking innovativeness and originality of setting. He directed Scottish Opera's 1988 production of *Candide,* conducted by John Mauceri.

John Mauceri has said that during the Scottish Opera production of Candide *you, as the director, brought a European sensibility to the work . . .*
Yes. I tried to some extent to restore it to Voltaire, and to make Europe at least the center of it. What had happened was that layer upon layer of rather vulgar New York vaudeville had replaced practically everything. In which case why bother to do *Candide* at all?

Bernstein referred many times to the fact that Candide *had been the "stone in his shoe," the work over which he had labored most.*

I got the impression that he felt we had delivered it, finally, to him. He said: "We've got it now." I didn't hear the later London recording [for DG in 1989] but I suspect it had that usual thing of being heavily burdened with famous people, and probably lost all the character that it had on the stage.

How do you react to opera stars performing works like Candide *or* West Side Story*?*

I think it's absolutely disastrous. But that has always been an aspect of Bernstein. Although he was happy to have these things done from time to time in the way that was most suitable to their genre, ultimately he just loved big-time showbiz and loved the big names in it. I think the thought of Kiri te Kanawa singing anything like *West Side Story* is completely absurd.

As a composer Bernstein aspired to be recognized for his serious pieces . . .

Yes, but I suspect that as a composer what he will be remembered for are these great American musicals. He is not a great "classical" composer of any importance at all, I would have thought. I'm not a musicologist, so I don't really know enough about it, but it has always struck me that what he will be remembered for are these extraordinarily memorable songs and ensemble numbers—"America" is one of the great musical episodes of the twentieth century. Of course, as always with New York-Jewish-Showbiz, and I speak as a Jew who can recognize it a mile off, there is a kind of sentimental, saccharine tosh that runs through everything. The sentiment of a piece like *West Side Story* is quite insufferable, I think. There is that song, "Somewhere," for example. You find yourself wanting to say: "No, Lenny, there isn't such a place." Whereas "America" or "Gee, Officer Krupke" are wonderful pieces. I think "Tonight" is also rather good.

Do you think Bernstein's tenure with the New York Philharmonic stopped him in his tracks as a composer?

It may be the case. But I think he *was* a great conductor—he is up there with the big ones—and he was promiscuously over-endowed as a musical talent. He maybe had too much of it, he could do anything he wanted to. Except write really serious, original music. Which, of course, is what he *most* wanted to do. This so often happens. On the other hand, composers like Irving Berlin were content to be what they were: great, memorable songwriters of the twentieth century. Then there is Gershwin, who meddled with the concert hall, but who is in the end remembered for those extraordinary pieces of Americana. The thing that always fascinates me is that all of America's "apple-pie" marching tunes were actually written by Jews. The tunes to which America falls in behind the leader, falls in behind the flag, so to speak, from "Appalachian Spring" and "Rodeo" to "Give My Regards to Broadway," and "Yankee-Doodle-Dandy" are all the work of Jews.

To return to Candide, *it has been said that one of the work's problems was that it was too literate a piece for the public. Would you agree?*
Perhaps the original version was, but I think all subsequent performances did their best to obliterate that. It's an uneasy mix, and I don't know why it was chosen in the first place. I think it perhaps has to do with that awe-struck thing that Americans have for European culture. But they couldn't resist turning it into vaudeville, and it is filled with the most appalling pieces of kitschy vulgarity, I think. It's also very hard to stitch together—it's actually a Frankenstein monster. It's absolutely pieces of dead bodies, that have been hitched together, and you keep on having to haul up the jacket and the shirt to hide the stitch-marks.

Many works of Bernstein, for example, A Quiet Place, *have been altered and moved around structurally by other people, and what is strange is that Bernstein does not seem to have minded.*
Yes, that is odd. It was as if he knew intuitively that they were

not pieces of great value. He wasn't a natural dramatist, I don't think. He does have great, sudden moments, but what is interesting is that the works that people are most likely to remember are in fact derivative from works of literature which are more memorable than they are.

A critic wrote recently that Bernstein was "irresistible and intolerable at one and the same time." Would you agree?

He was absolutely that. I remember in Scotland he turned up as the great, smoking celebrity at the very end of *Candide,* wearing all white, and shoved his tongue down everyone's throat. He rather mystified and repelled everyone. They were filled with awe and horror at the same time. And he was arrogant and noisy, and I think that everything that people said about "radical chic" was true of him. Of course, he was very intelligent and he had that sort of omnivorousness, which enabled him to pick up all sorts of references and cross-references, and which would have helped him, for example, to give those lectures at Harvard. But I find the Norton Lectures rather shallow. Those references to [Noam] Chomsky I think were all nonsense. In my opinion he really didn't understand what Chomsky was on about at all. He just got a sort of showbiz version of it, and people were rather staggered to hear the composer of *West Side Story* mention Chomsky.

How about The Young People's Concerts*? There he seems to have made an exceptional contribution.*

Oh, yes. I think he did have a way of getting things across. He was very charismatic, and tremendously energetic, and in some ways with younger people, people who weren't in competition with him, he was very generous.

To move on to Bernstein's relationship with the Viennese, does it seem strange to you that an American Jew should have enjoyed such adulation there?

I don't know how he overcame that. Possibly the sheer energy of his work, and, of course, he was an attractive character. But also, he presumably came slightly at an odd angle to the stupid Viennese expectations of a Jew. Presumably they were entirely modeled on sort of "Stürmer" pictures, and they expected there to be a ringleted, mittened Shylock, and then along came this leprechaun. The thing about anti-Semites of that sort is that they are so stupid that if they are confronted by someone who is in any way different from their stereotype, they assume that he must not be a member of it. I must say that I have to take many, many baths after I leave Vienna. It was in Vienna that I heard a rather irritated young Catholic student, who was railing on about the Holocaust, say to me: "You know, it was very exaggerated. It was two-and-a-half million, at the very most."

Tim Page wrote about Bernstein that only in Richard Wagner had the sublime and the silly been so inextricably yoked. Would you agree?
Yes, but both the sublime and the silly are in lower case with Bernstein. Whereas, in the case of Wagner they are in bold capitals. Wagner's silliness was really monstrous and absurd, but it is offset by the fact that he did write some of the greatest music ever. And that can't be said of Bernstein. Bernstein, I think, was a wonderful showbiz composer, whose talent rather inconveniently overflowed into other areas.

How much of Bernstein's popular success as a conductor do you think can be attributed to the fact that he was physically very attractive to his public?
He was handsome, certainly, and he had a certain mischievous quality. There was this feeling of a lovable, almost magical figure, who put everything he had into a performance. He seduced the orchestra, amused the audience—I suppose he was like a sort of musical Rumpelstiltskin.

In fifty years' time what do you think we will remember Bernstein for?
I think for the musicals, probably for *West Side Story* and perhaps little odd things like *On the Town.* I'm not even certain that

Candide will be remembered—it's such a weirdly fragmented piece. So much work has to be done by those who come afterwards to stitch it together. It really is always threatening to fall apart at the seams. Even if, according to Lenny's standards, we delivered what he had imagined it to be, I would think that no more than three months after we finished it, reconstituting what we made of it would have been very difficult. The fact that it may have worked had, I think, very largely to do with Anthony van Laast's choreography, to some extent my direction, and then John Wells's reworking of the text. That was perhaps over-emphasized, but for some reason Lenny put a great deal of trust in John Wells.

What is interesting is that since the Scottish Opera production, there appear to have been no further stagings of Candide.

No. And it doesn't belong just as a concert-recording [Bernstein's 1989 DG version]. There are a lot of nice, beguiling songs, but you don't know what they're about unless you see them in context. It's extremely thin as a piece of drama—it doesn't work. I began to feel desperate about it halfway through my directing it; endless bits of it seemed so silly and vulgar.

Do you think Hellman's original book, which has been so much maligned, was possibly better?

I don't think so, really. I think it was probably better than all of the stuff that got piled on later. Practically everyone in America seemed to have done something—I'm amazed that they didn't go down and ask the janitor for a bit! It is exactly what biologists mean when they use the term "chimera." It is one of these pieces patched together out of genetic fragments. It can, I think, in the hands of a fairly skilful director and choreographer hold together. But it is not really re-constitutable from its elements. The fact is that *Candide* very rarely rises above the quality of one or two of its individual songs, and a little bit of morbidly imaginative showbiz invention on the stage. I certainly myself wouldn't want to be remembered for anything I did in it; I perhaps made it a little more tasteful than it would otherwise have been, but it certainly is not something that, as it were, taxed my stagecraft.

It has been said that Bernstein brought a questionable taste to much of what he did, whether as composer or conductor. Would you agree?
Well, there is questionable taste throughout the whole of *Candide*; it's very vulgar, some of it. For example, "What a Day, What a Day for an Auto-da-fé" is a wonderful piece of music, but it is absolutely scandalous to have it in the middle of something which is actually to do with burning people to death. And it is not only filled with questionable moral taste, it is also filled with pieces of real Broadway shallowness—it's real garment-trade stuff.

Was there, do you think, the feeling that Bernstein was himself a somewhat vulgar personality?
Well, as I said, when he came to Scotland on the last two days, he tongue-kissed practically everyone in the cast. People were prepared to be enormously admiring, and then quite suddenly everyone was grasped by him. Being kissed by him was like an assault by a sort of combination of sandpaper and sea-anemones!

What do you think prompted this kind of extraordinary behavior?
I think as he went on he became personally and sexually, and I suspect musically and dramatically, incontinent.

Do you think that the entourage that went around with him wherever he went was damaging to him?
Well, I think he always had this world of an entourage, and he had this New York, Upper West Side showbiz thing—"Glitter and be Gay" is, I think, a rather admirable statement of his whole life. Or perhaps it should be "Glitter and be chic"!

VI
Great Performers on Bernstein

Jerry Hadley

Jerry Hadley made his operatic debut at the New York City Opera in 1979, as Arturo in *Lucia di Lammermoor,* subsequently joining the company and singing roles ranging from Massenet's Werther to Tom Rakewell in *The Rake's Progress.* He made his European debut at the Vienna State Opera in 1982 as Nemorino in *L'elisir d'amore* and has since appeared at major opera houses throughout the United States and Europe, including those of Chicago, San Fransisco, Hamburg, Munich, the Deutsche Oper, Berlin, Canadian Opera, Covent Garden, and Glyndebourne. He made his Metropolitan Opera debut as Des Grieux in *Manon Lescaut* and has since been a regular guest there. On the concert platform Hadley has worked with many major symphony orchestras and conductors and includes among his numerous recordings, for Deutsche Grammophon, Rodolpho in *La Bohème* and the Mozart Requiem under Bernstein, as well as the 1989 *Candide* (the title role) with the composer conducting.

Could we begin with your performance in Candide, *and the subsequent recording of the work with Bernstein?*

This was an epic story. In December of that year there was an incredibly virulent influenza epidemic, and we were scheduled to do concert performances of *Candide* at the Barbican after a week of rehearsal and then to go immediately into Abbey Road Studios and record the piece with Lenny. What happened was that the cast and Lenny began fighting a losing battle to stave off this awful 'flu. Throughout the week of rehearsals there were problems, with people getting ill, and then finally we got to the performance. June Anderson and Christa Ludwig and I were holding on with a wing and a prayer. Watching the video of that evening—thanks to the wonders of technology we were able to go back in the studio and re-dub a few bits here and there—there's a point after one of Candide's "Laments" when you see me in the frame of the picture and Lenny's hand reaching out to take my elbow at the end of the song, as if to say: "Good job, kid." Actually what Lenny was doing was preventing me from keeling over! I had finished a very long sustained note, and I had begun to black out and Lenny saw that, grabbed my elbow, and said: "Hang on, kid, hang on!" I sat down and fortunately we got through the evening.

The next evening both June [Anderson] and I had to cancel because we couldn't even rise from our sick-beds. We both felt wretched, not just because we were sick, but because this piece had always been—as Lenny put it—the stone in his shoe. This was the one chance, and I believe that somewhere in his mind he must have known it was his *last* chance, to do the piece. Also we had a terrific cast assembled. I suppose retrospect makes us all seem clairvoyant, but I have seen photographs of Lenny taken during that period, and I have watched the video that they made of that first performance, and, you know, the spectre of death was all over his face. I have not yet been able to sit through the whole video in one fell swoop, because basically you're watching this man who is beginning to die. What was remarkable was that as soon as he stepped onto the podium, there was nothing that could diminish the power and the spirit and the life force that was in that man.

Do you think there was some kind of transformation in Bernstein when he mounted the podium?

Lenny was fearless when he got on that podium. And he became a conduit for something that was life-changing, both for him and for everybody else. My experiences with Lenny were during the last six or so years of his life. I don't know what he was like as a young man, I can only infer or surmise what he was like by watching the many films of Lenny that were made. I know that when one first rehearsed with him, one would walk into the room and think: My God, this is Leonard Bernstein, this isn't just another guy waving a stick! And yet Lenny would go some way to shattering the myth—more than that—he would take the extra step to embrace both physically and figuratively everybody in the room, so that there was a spirit of collegiality. What was interesting was that in rehearsals Lenny was constantly asking for us to break our bounds. And then when he walked onto the podium for a performance, it was not so much a transformation as a focusing of all the energy that he had been putting out in hundreds of different directions during the rehearsal. Somehow it would all focus like a laser-beam.

When he died, Deutsche Grammophon asked many of us to write a little statement, in tribute, and I remember I said: "When Leonard Bernstein was on the podium and we as the performers were looking into his face, he was showing us a world that was totally without fear. When he conducted you were convinced that neither were the gates of heaven unattainable, nor the gates of hell inassailable." He was powerful, and you believed that you could do anything. And he never used his position or his power at anyone else's expense. If anything, Lenny was willing to go the extra mile to assure a performer, be it a world-class superstar or a beginner. He treated everybody equally, he met them where they were, and he drew out of each of those people the best that they could summon up at that particular point in their lives. Lenny truly made love to everybody, and I don't mean that in a sordid sense at all. He really partnered the people who were performing with him, and, unlike many conductors, if something began to go awry, Lenny would never let anybody hang out there on the yard-arm by themselves. If things were going wrong, Lenny would jump right into the fray with whatever was happening, and either sink with it or rescue it. But he never abandoned anybody in the midst of a performance.

To return to Candide *for a moment, do you think that when Bernstein made the 1989 recording he felt that the work's time had finally come?*

Well, I know that many times during the course of those rehearsals he referred to *Candide* as the child that needed special attention, the piece that he had really labored with. I personally think that *Candide* is the most fitting legacy that Leonard Bernstein could have. It's so much an extension of everything he was—it's eclectic, it's witty, it's profound, it's irreverent, it's tongue-in-cheek one moment and innocent and full of childlike wonder the next. I have often thought that *Candide*'s biggest problem was that it was almost too literate a piece for the average audience. In those rehearsals I think Lenny was really anxious for the piece to finally be the way he had conceived it, with a cast that he had chosen.

The version used for the DG recording seems to have been largely that prepared by John Mauceri for the Scottish Opera production . . .

It's my understanding that the Scottish Opera version of *Candide*, which was a real labor of love for John Mauceri, went a long way toward crystallizing Lenny's ideas of how the piece ought to go. I know about the long and painstaking process that John underwent in order to bring that piece to the stage. So I think that in many ways John Mauceri is the unsung hero of *Candide*'s ultimate success.

Bernstein told Edward Seckerson in an interview in 1989 that he felt that he should spend whatever time was left to him composing, and that he was really not needed for another Ring *cycle or another* Magic Flute.

Bernstein with Lukas Foss and Koussevitzky, The Berkshire Music Center, 1940. (Whitestone Photo/Heinz H. Weissenstein)

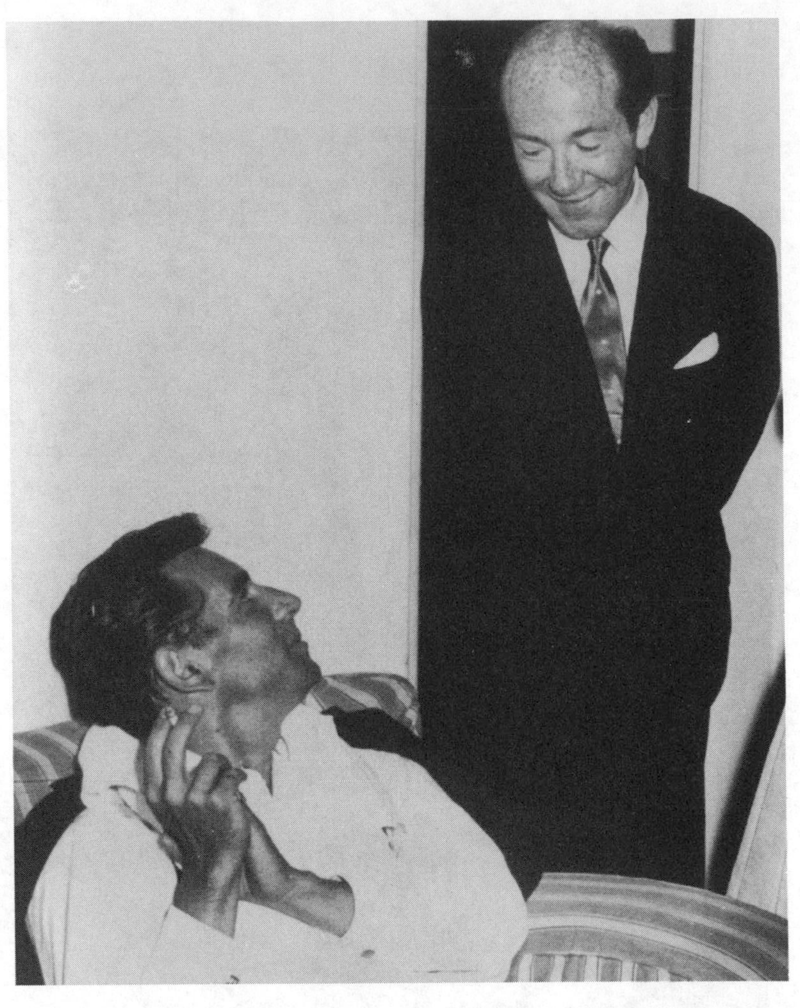

(Above) Bernstein and David Diamond following a performance of Diamond's *Romeo and Juliet,* Florence, May 1955. (Levi, Firenze)

(Above, right) With Stravinsky in Venice, 1959. "Neither Craft nor Stravinsky liked him. Stravinsky...didn't like the way he conducted the *Sacre [Le Sacre du Printemps]."* (Sony Classical)

(Below, right) Bernstein, Boulez, and Barenboim, celebrating the occasion of Bernstein's fifty-fifth birthday (August 25, 1973, Edinburgh). *Back row, far left,* Paul Myers. *Seated,* Janet Osborn, Felicia, Bernstein, Boulez, Barenboim, Jacqueline du Pré. (CBS/Sony Classical)

(Above) Bernstein with John Mauceri after a performance of *Mass* in Vienna, 1973. "I think Lenny was terrified that his music might be boring—I think he was more afraid of that than anything else...." (Heinz Hasch)

(Above, right) At Juilliard, 1979. "Lenny was happiest when working with young people; first and foremost he was a teacher." (1979 Peter Schaff)

(Below, right) *Candide* at Scottish Opera, May 1988. John Mauceri with Jonathan Miller and John Wells. (Eric Thorburn)

Bernstein with Jerry Hadley (and Angelina Réaux) during the concert performances and recording for Deutsche Grammophon of *La Bohème*, Rome, 1986. "I think Lenny saw *La Bohème* as an elegy to youth—his youth." (DG/Henry Grossman)

Recording with Rostropovich: the Schumann Concerto and Bloch's *Schelomo* with the Orchestre Nationale de France, Paris, 1976. (Eric Brissaud)

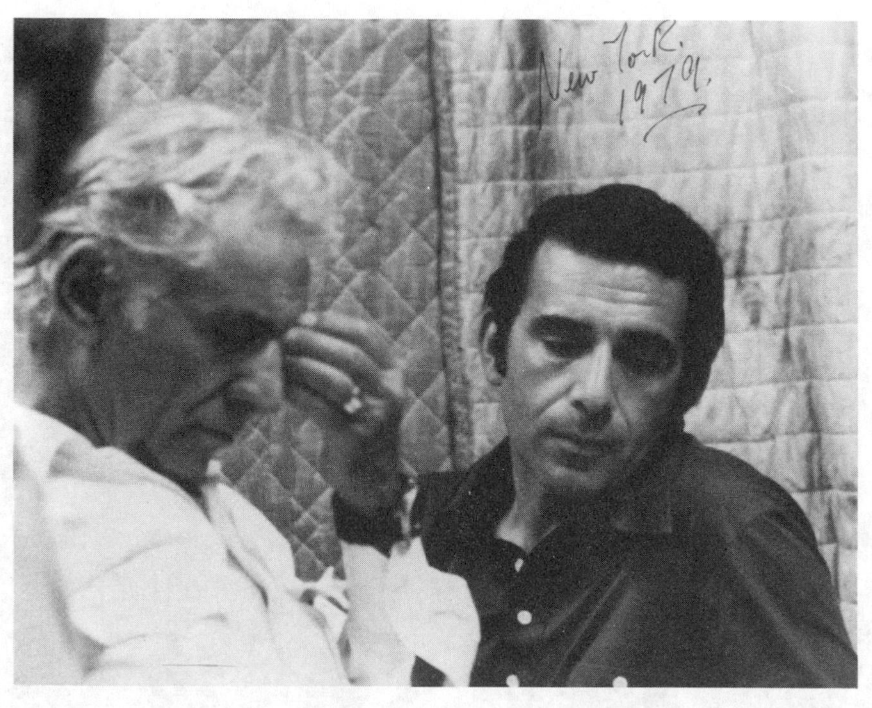

With Rodney Friend (at the time Concertmaster of the New York Philharmonic), New York, 1979. (Courtesy of Rodney Friend)

Bernstein believed Christa Ludwig to be one of the finest Mahler exponents of the day. (DG/S. Lauterwasser)

(Above) Carol Lawrence in the original production of *West Side Story*, 1957, in the improvised "I Feel Pretty" scene. "The first time the world was going to see us, I was being asked to improvise...in a Jerome Robbins-choreographed show." (Fred Abeles/Courtesy of Carol Lawrence)

(Facing) The Balcony scene in the original production: Carol Lawrence and Larry Kert. (Fred Abeles/Courtesy of Carol Lawrence)

West Side Story 1985, with the composer conducting. An all-star operatic cast was used for the Deutsche Grammophon recording. (DG/Susesch Bayat)

(Facing) Frederica Von Stade as Claire in *On the Town,* performed and recorded by DG at the Barbican, London, 1992. (DG/Clive Barda)

With Aaron Copland. "It is possible that some form of stage music will prove to be Bernstein's finest achievement...." (DG/Walter H. Scott)

I remember many conversations with Lenny late at night, after rehearsals, and when he was in his cups he would start bemoaning the fact that he was a failure as a composer. I remember his once saying: "My God, by the time Mozart was twenty-five he had written this. By the time Mahler was forty he had written that." On another occasion, I remember a young Bernstein acolyte in Munich—a young German fellow—extolling Lenny's virtues as a conductor, and Lenny saying: "You don't understand. This is my problem: When anybody says that I'm a great conductor, I think—no, I'm not a conductor, I'm a composer. When somebody calls me a great composer, I think—no, I'm not a composer, I'm a pianist. When somebody calls me a great pianist, I think—no, I'm not a pianist, I'm a teacher." So, with Lenny it was an embarrassment of riches.

Bernstein on one occasion told a member of the Israel Philharmonic that he did not want to be remembered as the composer of West Side Story *but rather for one of his serious compositions.*

It's funny you should say that, because I know a similar story. I remember being at a party at the American Embassy in Rome, the night of the concert performances of *Bohème* that we did at Santa Cecilia. Lenny said very much the same thing. I think it was something of an obsession with him. The ambassador was saying how he had always loved *West Side Story* and Lenny said: "Yes, but do you know how depressing it is for me to think that I'm going to be remembered only as the man who wrote *West Side Story*?" One of my colleagues, who had been involved in *A Quiet Place* and didn't really like the piece, turned to me and said: "Well, it's better than being remembered as the man who wrote *A Quiet Place!*" I do think that he wanted to be remembered for something other than *West Side Story*, not that he didn't love the work, but I think that he would have preferred to be remembered as the composer of, perhaps, *Candide* or the composer of a great symphony. This is the bane of the existence of a lot of people in the classical music world who gain widespread popularity. There is nothing wrong with being widely acclaimed, but in Bernstein's case I think that in the mind of so much of the public he was associated with *On the Town*, with *West Side Story*, with the show things that he did, that there was a large contingent that wasn't interested in hearing the *Chichester Psalms* or *Mass* or whatever, because it didn't fit their preconceived notion of what Leonard Bernstein was.

It's too soon after his death to really make a statement, but I think that when there is a reassessment twenty years from now, *West Side Story* will probably be an important part, but I don't think it will be the whole picture. I think that as a composition *Candide* is just as important, if not more important, and I don't say that just because I was involved in it. And there are the other works, he will not be remembered simply as a one-piece composer. Also I think that Lenny's role as an educator and his contributions to our appreciation in this part of the century of Mahler cannot be lightly dismissed. A subject that I hold very near and dear to my heart is the fact that Lenny, for my generation of Americans, invented music. And one of the things that was so surprising to me when he died, was the void that I felt. I was very surprised that I felt it as strongly as I did. I have worked with other people, equally as famous as Lenny, who passed on, and you would feel bad about it, or perhaps feel that it was in some way tragic. But I really felt *miserable* about Lenny's death. The only other time in my life that I felt that bad about a public person dying was when the Kennedy brothers were assassinated. It was the same feeling. Lenny had become such a part of our consciousness—for American musicians, particularly—that we took him for granted.

Do you think if less of his energy had been consumed by his public persona, his public life, that he could have given more time to writing?

I think if one ever spent any time with Lenny, one was left with the distinct impression that this was someone who really loved humanity in a way that most of us don't. I think that his public life did consume vast amounts of his energy, but on the other hand, from what I can see, it was that public life which energized him. It's a paradox. There are reams and reams of material by Bernstein's detractors, taking him to task for certain choices that he made in his personal life, for certain excesses, and the implication is that had he not lived his life a certain way he would have had more time to write. The implication is that those extravagances made him undisciplined, but I take issue with that, because of the many words I could use to describe Leonard Bernstein, undisciplined is not one of them. Whatever his excesses were, they seemed a logical extension of his willingness to go farther. In a piece of music he would be willing to go farther than almost anybody to investigate all of the possibilities.

In terms of the choice of conducting versus composing, there is an anecdote by a New York friend of Lenny's—she must be in her eighties now—which she told me shortly after he died. I was bemoaning the fact that he didn't have more time to write, and she said to me: "We used to make a joke about Lenny. We thought we knew how to get Leonard Bernstein to write the Great American Opera. All we had to do was book Carnegie Hall and fill it with Leonard Bernstein fans, and have a table with manuscript paper and a piano on the stage. Lenny would sit down, write a phrase of music and then receive a standing ovation. Then he would go on to the next phrase." She said that with great affection for Lenny; it was not meant as a criticism. Her point was, of course, that he couldn't *not* conduct, he couldn't *not* be in front of the public.

Was there something special attached to a live Bernstein performance?
The sum of the parts in a Leonard Bernstein performance was an astounding thing. It wasn't just an aural experience. I can

remember a Sibelius Second Symphony that I heard him conduct in Vienna in about 1986. There were two performances with the Vienna Philharmonic, which were being recorded, and the performances were utterly phenomenal, they were life-changing experiences. I rushed out and bought the recording, and I listened to it, trying to recapture that feeling, and I couldn't. It's not that the performance was bad on that recording at all, but there was something missing. When Lenny walked onto the podium and began to conduct, it was as if he embraced all of the fears and all of the hopes of everybody in the room. He took us on a ride that we were not necessarily prepared to go on, but at the end of the performance we were different people.

The things that he was often taken to task for—the dancing, the flamboyance on the podium, and what was often called self-indulgence—I have to say that we as performers did not feel. His willingness to lay himself open like that gave us the courage to do it ourselves, to go for things, and in many cases it simply made us laugh at ourselves, so that we would not take ourselves too seriously. And that helped us focus on the music. I'm not sure that Lenny took himself all that seriously, but I think he took his work deadly seriously.

As a singer, how did you find his tempi?

Lenny's tempi were slow, but they were not unsingable, because Lenny always had a sense of rhythm, and a sense of phrase movement. There are very few people, Karajan included, who could take a tempo and stretch it the way Lenny could. I'll give you an example. With *Candide* I had to go into the recording studio nearly eight months later, and track my part over the orchestral tracks that Lenny had laid down. And when he laid down those tracks he was a very, very sick man, so those tempi were slow. But they were not erratic. All of those songs had a structure to them. They may have been slower than I remembered them, but they were not unsingable. All it meant was that one had to find a different way to phrase, and perhaps take a breath here and a breath there that you wouldn't normally take. It was never unmusical.

I remember when we recorded *La Bohème* together in Rome in 1986. Lenny believed in pushing us beyond what we thought we were capable of, and we many times went over the top, and we many times got pulled out of our best stance as performers, in an effort to do what we thought Lenny wanted. At the end of the day, I'm not sure that that *Bohème was* what Lenny wanted, and it certainly didn't end up being what most of us wanted. There was something that didn't work about it. I have never sung "Che gelida manina" that slowly in my life. *But*—I have never found it easier to sing. For one thing, there was a lot more time to think and, in a strange way, more time to breathe. And I don't think that what went wrong in that recording for me should be looked upon as Lenny's fault. I think that I often misinterpreted what Lenny was after. I think Lenny saw *La Bohème* as an elegy to youth—his youth. One of the problems was that we, the cast, were trying to do a *Bohème* of the here and now, whereas Lenny was almost making it into a tone-poem about his own bohemianisms as a young musician.

Do you think he fits more into the mould of a nineteenth-century interpreter, rather than a musician of the more literal modern approach?
Lenny believed that it was his duty as a musician to start with what was on the page, to fully comprehend and to fully honour what was on the page. But he also believed that because he was an individual with his own history and his own feelings and his own hopes, that it was his duty to meld those things with what the composer had written, and to bring the music to life in a way that was more than just the right notes in the right place at the right time. In the hands of somebody who is undisciplined, that approach can pull music out of shape and make it a self-indulgent exercise. I think the reason that most of the time it

worked for Lenny was because he *was* disciplined. If he chose to do something differently, he knew why he was doing it. I think that his intellect and his training and his impeccable preparation on anything he did, allowed him to be spontaneous and, sometimes, serendipitous.

To return to opera for a moment, it seems strange that he never conducted works like Don Giovanni *or* The Magic Flute . . .
I remember whem we were doing the Mozart Requiem together in 1988, Lenny said to me on many occasions how he loved *Don Giovanni* and I remember he said: "I'm going to have to do that piece some day." But that's the only thing that I can ever remember his saying about that. Can you imagine what that would have been like? Something else that I regretted that he never got around to, from a purely selfish point of view, was the Britten War Requiem. He had broached that with me before he died. I think that would also have been a memorable experience. But when you consider the wealth of repertoire out there, he did manage to get through quite a bit of it.

It has been said that when Bernstein was accepted in Vienna, his reputation in America changed overnight, and he was treated with a new deference. Would you agree?
Yes, but I don't think this type of thing is limited to America. Sometimes one has to go off somewhere else and be validated by another country, another public. I suppose it's like the prophet in his own land. I know that in my case I had worked very hard for six or seven years in America, and was very well known in my profession there. Then I had my first European engagement at the Vienna *Staatsoper* as Nemorino. It was a big success, and all of a sudden when I came back to the States I was treated with enormous deference. Now, with somebody like Lenny, who was already a person of stature, that would have been multiplied several times over. Here was an American Jew—of all things—who had gone over to Vienna and conquered that city. And not

just conquered it; he really made it his own. Those people would have marched into hell for him.

It has been commented on many times how strange it is that Bernstein should have had a love affair with what is reputed to be an anti-Semitic city and orchestra.

You know what, he called them on it all the time. He flung it in their faces. And they loved him for it. It was as if he was saying: "Look, I know and you know what despicable bastards you can be. But I don't care. Let's make music."

Some musicians have said that Bernstein was not always easy to follow in terms of technical clarity. What was your experience?

I think that with Lenny you had to look at the whole experience. If you rehearsed with him, you understood what he was trying to do. He wasn't necessarily bound by saying: "Here's one, here's two, here's three." Somehow his body would pulsate with rhythm, and there was something about his focus that told you what was going to happen, without even indicating it. Also, Lenny did trust the people with whom he was performing, and sometimes he was content to just let them sing, or play. In all the things I ever did with him I can only remember one instance where I didn't know what the hell was going on, and that was during the *Bohème* performances. There was one place in the first act, right after Mimi's entrance, where I didn't know what was happening. I didn't really understand what Lenny was doing. But that may have been my fault.

How did you find Bernstein's interpretation of Mozart, more specifically the Requiem?

It was interesting, it was the second Mozart Requiem recording that I had done in about four years. I did one with Robert Shaw and the Atlanta Symphony, and Robert, while not pedantic, was very careful to do things the way he thought Mozart had written them, the way Mozart wanted (whatever that means).

Lenny decided that he was going to perform a Requiem Mass that was indeed a cry to grant rest to the souls of the dead. I have never heard the "Lachrymosa" sung with such passion as with Lenny, it was just hair-raising. He did some things that would have set the Holy Temple of Mozart Scholars spinning like dervishes. With the last "Quia pius es" in the Requiem, he had the chorus attack the "es" almost with a *sforzando,* and then he did the longest and most breathtaking *diminuendo* that you have ever heard in your life—they must have held it for ten or fifteen seconds. I remember at the first performance—we did it in a little Baroque church in Diessen, in what was then West Germany—he did the attack on that last note and then took it down to nothing, and just as the sound died away the church bells began to ring. Bong! Bong! It was almost as if God was answering, and the audience could not bring themselves to applaud. They sat in stunned silence. Lenny put down his baton and we all turned and walked off, and there was not one iota of applause. The orchestra and chorus sat there for a good three minutes, and then the audience began to applaud and we all went on and took our bows. I've never had that happen in a Mozart Requiem before. And with his version of something like the "Benedictus," you have never heard it so fast, but it was incredibly positive and uplifting. I found it a very unusual interpretation of the Mozart Requiem; it was tremendously passionate, not that the music isn't passionate already, but somehow there was a suffering and a joy that Lenny injected into it that is rarely heard in the piece.

How do you place Bernstein as a conductor?

Lenny is not categorizable. There's simply Bernstein. Period! I know that he revered Koussevitzky—Lenny wore Koussevitzky's cuff-links until the day he died. He would wear them for every concert, and he had a ritual—he would always kiss the cuff-links before he went out on stage. I heard him talk about every great conductor of the previous generation whom he knew, and, knowing Lenny—he was such a sponge—he probably absorbed something from everybody. But I think his genius, and the thing that we all wish we could do, and that we all strive to do, is that he did that and then made it his own. Maybe one of the reasons we have felt his loss so severely is that there *isn't* anyone like him. I suppose of all the Bernstein disciples, Michael Tilson Thomas is probably the one who is the most willing to go for things in the way Lenny did, but Michael is his own person too. There's never going to be another Leonard Bernstein. Although Lenny in many ways was a real product of his time, I think he also harkened back to a simpler and more innocent world. There was this child-like wonder with which he approached everything. And I think the quality of Bernstein's genius, and the force of his personality and intellect, is something which is very rare.

Many critics have suggested that Bernstein as a composer is derivative. Would you agree?

I don't think you can ever accuse Lenny of being derivative—you can certainly say he is evocative, but never derivative. And isn't every composer evocative? I think that is one of the things that is so ludicrous about much of the criticism that is heaped upon composers today. It's almost as if the critical community wants the composer to operate in a vacuum. And they can't—how can they do that? If you listen to Lenny's music, part of his genius was that he had it all at his fingertips and he could summon it up at will. If you look at *Candide* he uses almost every compositional style that you can imagine. He even uses a twelve-tone row, in the "boredom" number, but it still sounds like Bernstein. And apropos of twelve-tone music, I remember once hearing Lenny say: "Any asshole can write a twelve-tone row; it takes a composer to write a melody."

Do you see Bernstein as one of the first of the more democratic conductors, as opposed to the autocratic old school?

That is absolutely the way that I perceived it. But with Lenny no-one was ever in doubt as to who was in charge, and somebody has to be. At the end of the day somebody has to take responsibility for giving the piece a point of view. And in my experience, those conductors who enjoy power for the sake of power often don't have a point of view about the piece. With Lenny, while he would involve other people in the whole process of the interpretation, he would never relinquish responsibility for it. And when he would get angry or frustrated at somebody, he would never make a personal attack on them. He may have been capable of it, but I never heard it. I remember his getting angry when people weren't paying attention, but as quickly as the anger would flare up, it would dissipate as soon as he had regained their attention and he would say: "Right, now let's have fun." Also I think that because of his whole approach he allowed large performing forces to behave like chamber musicians. And that was a great service.

Bernstein spoke often of "re-composing" a piece of music when performing it; was this something that he ever discussed with you?

I never heard him actually say that, but I experienced that with him on more occasions than I can remember. Performers are always spoken about as being "creative," but the kind of creativity that a composer has is something that we performers don't have. Essentially what we do is to take something that somebody else has plucked out of the air, so to speak, and bring it to life. We don't just parrot it back; we hopefully, in some way, eke out the humanity of the piece. To hear Bernstein describe it in terms of "re-composing" makes a lot of sense, because he had the perspective of both performer and composer. And I think he did do this—he certainly re-composed the Mozart Requiem for me. Also I can think of numerous symphonies that he brought to life in a new way. Look at what he did for Carl Nielsen's music. Lenny was not an exclusive musician; he was totally inclusive. He never in my experience bad-mouthed *any* kind of music. Lenny could take the most mundane piece of music and somehow find the merit in it.

How do you think he felt toward his own music? Was there a sense of ambivalence?
I have a story about that. Just this past May in New York we were doing a gala of Sondheim music and Stephen was there. He was fidgeting around at the dress rehearsal, and I said to him: "Is this really hard for you to sit and listen to one composition of yours after another?" He said: "Yes, it's torture. The only person I ever knew who loved every note he wrote was Lenny." He said that when they were writing *West Side Story,* for example, Lenny would play something he had written the night before and say: "Isn't that great?" I think also maybe Lenny enjoyed his own music most when he was performing it.

What do you think Bernstein will be remembered for as a conductor?
I think he will be remembered as a great interpreter of Mahler, and of Beethoven. And I think he will be remembered as a great teacher. That was so much a part of his conducting, too. And when I say that Lenny was a teacher, I should add that you never felt he had an axe to grind. Often one will go to a concert with a conductor whom one feels has a particular point to make, and one leaves the concert feeling that the conductor has condescended to share a little morsel that we, as the audience, couldn't possibly understand. Lenny, on the other hand, always presented the music as if to say: "Come here. I've got something great to show you!" I feel that in some ways that was his greatest gift, that communication, and in many cases he opened the eyes of the blind. In terms of the future, I think that after many other names have faded away Lenny's is going to be remembered. I think that in two hundred years' time, when the great conductors of the twentieth century are talked about, he is going to be up there in the first row. What is amazing to me is how

universal his appeal was. Herbert von Karajan was a great conductor, by anybody's measure, but he was not a beloved human being. Lenny was beloved wherever he went. I think in many ways, in spite of all his personal excesses, he represented as a musician and as a human being, the best of what America has to offer. And I feel sometimes that to use the past tense about him isn't quite accurate. During a recent *Candide* performance [August 1992 in London] Della Jones looked at me and said: "Lenny's here tonight, isn't he?" And I must say, when I sang "It Must Be So," which starts out with the words "My world is dust now, for all I loved is dead," that rang inside me in a way that it hasn't done before. It was a very special personal moment for me. But the reason I say that it's inaccurate to use the past tense when you're speaking about Bernstein is that anybody who ever worked with him, who had their eyes open even a little bit, couldn't possibly go away from that experience able to make music the way they did before. I have felt many times when I'm performing something that Lenny is looking over my shoulder, saying: "Come on, kid! Go for it." So I'll tend to think of him in the present tense.

Mstislav Rostropovich

Mstislav Rostropovich was born in Baku on March 27, 1927. The great Soviet cellist (also a conductor and pianist) entered the Moscow Conservatory in 1943, studying cello with Semyon Kozolupov and composition with Shostakovich and Vissaryon Shebalin. In the late 1940s he won competitions in Moscow, Prague, and Budapest and there followed one of the most important musical careers of our time. Rostropovich has had many major works written for him, by such composers as Prokofiev, Shostakovich, Khachaturian, Panufnik, and Britten. After making his debut as a conductor in 1961, he conducted *Eugene Onegin* at the Bol'shoy Theatre, Moscow in 1968, with his wife, Galina Vishnevskaya, singing Tatyana. Since 1977 he has been Music Director of the National Symphony Orchestra in Washington, D.C. In a career studded with prizes and awards, Rostropovich has been the recipient of the Stalin Prize (1951) and the Royal Philharmonic Society's gold medal (1970), as well as an honorary degree of Mus. D. from Cambridge University (1975).

Do you remember the first time you worked with Bernstein?

Of course, yes, but before that we met first at each other's concerts, sometimes backstage or sometimes with friends at a restaurant. I felt that I knew Lenny all my life even before we played music together. Then I played with him two concerts in Paris at the *Théâtre de Champs-Élysées,* the Schumann Concerto and

Schelomo of Bloch. Also we made a recording [with the Orchestre Nationale de France, in 1976] of these two pieces. I remember I came there before Lenny, to warm up a little bit with my cello, and I was surprised because I saw on the conductor's podium a big mattress! So I asked: "What is happening here—is Lenny coming here to sleep?" And the producer said: "No, it's because he jumps. We use a mattress, otherwise he makes a noise."

Also, before that, I played with him the Schumann Concerto with the New York Philharmonic. I remember before the recapitulation in the first movement of the Schumann—on that famous note—I savored the music much more than a normal cellist, and in the rehearsal with the Philharmonic Lenny stopped the orchestra and embraced me, in tears.

How did you find his ideas about tempi and dynamics?
We were very much in agreement on the interpretation, I remember.

When Murray Perahia played the Schumann Piano Concerto with Bernstein, he said he found the tempi much too slow.
You know, I did not feel that in *my* participation with him, but I felt it sometimes when he was conducting *alone*, in the last period. For example, the Tchaikovsky *Pathétique* Symphony, in my opinion that was much slower than it needed to be. But you know, what is slow, what is fast, these are more questions for music critics than musicians. I will tell you why: music critics, they will tell you immediately: That's a good tempo, or that's a bad tempo. *I* tell you: That's not *my* tempo, but *also* a good tempo. It's another point of view. I think that if Lenny Bernstein, such a great musician, changed his tempo according to his feeling, that was all right. I don't think he became older, but rather more experienced. For me, as a conductor, it was the opposite—when I started conducting, thirty years ago, at my first performance I did every composition slower than traditionally. And why—I had such joy from the *sound* of the orches-

tra, I made every note longer. Now I am more experienced and I make the music a little bit faster, especially for recordings. On the concert stage I have my personality, but on a record it is like canned food—not so many vitamins! The presence of the personality, also electricity is gone. And maybe it was the same for Lenny. A live concert, for all of us, is *much* more important than a recording, much more. And especially for someone like a Bernstein. Bernstein had such a great combination of intelligence and *spontaneous* temperament, spontaneous interpretation. In recording he had maybe a more balanced point of view of an interpretation.

Bernstein has perhaps received more adulation from the public than any other conductor this century. Why do you think this is?
I think he appealed more to the public in his art than other musicians. He always made his music for the public. For instance I played many times with George Szell—he was a genius musician, a very great conductor—but he isolated himself a little bit from the public. He made his speech, but it was not so important for him how it was received. When Lenny conducted, he made his speech *for the people,* and then he would want to see how it was received.

What do you think he did best in terms of repertoire?
I know many of his recordings, and each recording has some very great quality. But I think that Mahler is one of the best. As for my own recordings with him, the Schumann I like, but with Bloch's *Schelomo,* there he made for me an enormous impression as a musician and as a man. You know, sometimes Lenny, with his temperament, would not always avoid exaggeration, sometimes he would be a little extreme. I worried a little bit, when we came together for *Schelomo*; I thought: Now he will make opera, he will make big theatre out of that. But, you know, the opposite happened. His understanding was so deep, he felt so deeply inside this music. He was much less extravagant as a conductor

in *Schelomo* than in Schumann. It was very interesting. With *Schelomo* the work affected him very deeply, very emotionally. After this recording, I told the record companies I did not want to make in my life another recording of *Schelomo,* and that is very rare. I did not make a record of the Sonatas of Shostakovich again, because I made this recording with Shostakovich, and I did not make a record of the *Arpeggione* Sonata of Schubert after my recording with Benjamin Britten, because nobody plays Schubert better on the piano than Benjamin Britten.

To move on to the compositions, Paul Myers has said that Bernstein had a burning ambition to produce a serious masterpiece, and that maybe it didn't come off. Would you say this is true?

Lenny, when he dedicated himself to composing, or conducting, or the piano, would try absolutely his best in each different field. I remember when he did *1600 Pennsylvania Avenue,* he was very serious, very dedicated. Often he told me: "I must stop conducting, I must stop playing the piano, I must absolutely be dedicated to writing music, to composition." Why I know about his perfection is because I played with him not just as a soloist, but as a piano partner—we played the C major Sonata of Mozart for Four Hands! I worked for this performance for two months, because Mozart on the piano is very difficult, especially for cellists! Lenny worked for two hours, but his results were much better than mine. I played the bass register, as a cellist, and Lenny played with more virtuosity in the high register. In the first rehearsal Lenny told me: "Slava, you know normally you would command the pedals, but maybe because I am so nervous, I can pedal—sometimes in a passage I might use more pedal than you would like." I told him: "Yes, Lenny, whatever you like."

Then when we sat on the piano bench for the concert at the Kennedy Center, Lenny not only took the pedal but he took three-quarters of the seat. After that, I said to Lenny: "Why when *you* play a passage you use more pedal, when *I* play the same passage an octave lower you use *less* pedal?" But we had a very good time on this work!

As a composer what do you see as his greatest achievement?
First of all I love his *Meditation* for Cello, and I would be happy to play this work all my life. I think he had great possibility as a composer. I also think that without him the United States could not have existed musically. Because he is a *portrait* of United States music. His Suite from *On the Waterfront* I have conducted many, many times, and this music *smells* of the United States. But it is a good smell! And I think everyone takes from Lenny's compositions what he likes personally. For example, Michael Tilson Thomas recently conducted *On the Town* in London, and Clive Gillinson [manager of the LSO] told me it was an enormous success. I was very happy for Lenny. I will say that when Lenny tried to become a deep composer, like Mahler or, for example, Beethoven, then it was perhaps not so successful for his composition. But when he tried to make a portrait of himself, his emotion and temperament, then he was a good composer. He loved his listeners, his public, and when he wrote for the theatre, then he could explain everything to the public. That is why Lenny, when he wrote a musical, was absolutely clear about what he was saying to the public. And that is why also with me I like to conduct opera, because opera helps me, as a musician, explain the story to the people. With just music, some people understand, some people don't understand, but opera or theatre is understandable for everybody.

It has been said that Bernstein was a transition figure between autocratic conductors like Szell and Reiner, and the younger more democratic generation. How do you see him? Where do you think he belongs?
Well, of course, he was certainly not of the same school as Szell or Reiner. First of all, Lenny's relationship with the orchestra was completely different from any of these people. He was

much more friendly, he spoke with the orchestra as a colleague. I didn't know Reiner, I never played with him, but Szell was very hard with the orchestra. That's the first difference. Second, I think that what for me is very sad is that Lenny was with the New York Philharmonic for only eleven years—I think that's not enough. If you talk about Bernstein as a conductor, I think he was for too long a period without his own orchestra. As such a great musician he should not have been only a guest conductor. His life as Music Director was only eleven years—that is too short. And I think that it is also a tragedy for the New York Philharmonic that he came away so early. If you think about Solti—I played with him many times—or Fritz Reiner or George Szell, these are three people who created great orchestras and created orchestras like their musical portraits. Solti changed the Chicago orchestra a little bit after Reiner, to make it more to his taste. And Cleveland, which became such a famous orchestra, should be grateful first of all to Szell.

Do you think the New York Philharmonic regretted letting Bernstein go? Ostensibly he left in order to compose.

Of course that is all very long ago, but I think if an orchestra wants a great conductor or a great musician, it must be willing to make a compromise. And on this occasion, with Bernstein, they could have said: "We understand your composition needs, your composition successes are our successes; we will give you for two years only five or six weeks of conducting instead of ten weeks." And they ought not to have burned his bridge as Music Director. Of course, Lenny did not change many people in this orchestra while he was there; this is a dirty job for a Music Director, but sometimes you must do it if you want to keep a great orchestra.

Some commentators have suggested that as Bernstein became older he took on too many projects, more perhaps than he could manage. Would you agree?

Yes. I think he maybe took on too much. Also I told him a

thousand times: "Lenny, stop smoking." I tried many times to speak to him about that. I begged him so many times. People tell me that after one of the last concerts that he did, in Prague, with the Ninth Symphony of Beethoven, he said to them: "Please, give me one cigarette." And that was not so long before he died. When Lenny died I must say that it was a great personal loss to me; he and Felicia, they were both very dear friends to me. And of course it was a very great loss to music; Lenny can never be replaced.

Christa Ludwig

The mezzo-soprano Christa Ludwig was born in Berlin and made her debut at Frankfurt am Main as Orlovsky (*Die Fledermaus*) in 1946. In 1955 she first appeared at the Salzburg Festival, in *Die Zauberflöte* and *Ariadne auf Naxos,* and the same year was engaged at the Vienna *Staatsoper,* where she has since sung regularly. She made her debut at the Metropolitan Opera in the 1959–60 season, and at Covent Garden in 1969. She has also appeared at Hamburg, Munich, Milan, Rome, and Chicago. She sings all the major mezzo roles and also some in the dramatic soprano repertory, notably Leonora in *Fidelio,* Lady Macbeth and the Marschallin (*Der Rosenkavalier*), the latter recorded with Bernstein and the Vienna *Staatsoper* in 1971. She is a renowned interpreter of Mahler, particularly his *Das Lied von der Erde,* recorded with Bernstein and the Israel Philharmonic in 1972, and has recorded *Des Knaben Wunderhorn* with Bernstein in the versions for both orchestra and piano. Ludwig has worked with, among other great conductors, Karajan, Böhm, Klemperer, and Solti, and worked regularly with Bernstein from the 1960s until his death in 1990. She played the part of the Old Lady in his final 1989 recording of *Candide.*

Your first work with Bernstein appears to have been in Rosenkavalier *in 1968. Could you describe this performance?*

Yes. I think my first meeting with Bernstein was during the period of the Six Day War between Israel and Egypt. First, I think, I did the Mahler Second with him and the Vienna Philharmonic and then we did *Rosenkavalier.* And when you saw the cast of *Rosenkavalier*—a black Sophie,[1] then a Welsh girl, Gwyneth Jones, as Rosenkavalier, myself playing the Marschallin, and I am more German than Viennese, an American Jew conducting—well, everybody said it was impossible. The only Viennese in the cast was Walter Berry, and he is no basso, he is a baritone! So. But Bernstein made it great. You know, Bernstein may have been American, but some of his roots were in Europe. He knew how to make this bitter-sweet three-quarter bar, he understood how to do that. And that was one reason why he could make *Rosenkavalier* so wonderful. Bernstein's upbringing and education may have been American, but even so his education still came from Europe. I often thought that in his way of making music he was more Russian than American.

Could you comment on Bernstein's extraordinary success in Vienna? His performances of Mahler, particularly, seemed to make a big impression there . . .
Yes, but also his Brahms and Beethoven. He had a big success in Vienna in any case. Of course, there was a certain time when the opposing party—I would say Karajan's party—was against him, but you have always this situation: you have Tebaldi and Callas, you have Jesus and Judas, you have always two sides who are against each other. Sometimes he [Bernstein] got very, very bad reviews. But you know this could be the same with Karajan. In Vienna there was a time when he could almost never get a good review. And the same was with Karl Böhm. I think this is almost a fashion—if somebody has a great success, they will get a bad review!

What is interesting is why the Viennese should take to their hearts a conductor who is not only an American, but also a Jew.
I don't think he was considered in the first moment as a Jew, not at all. He was *the* beloved conductor, suddenly, because I think

he brought a special joy to his music. You know, the Viennese, like the Germans, are very, very serious about music, really they are serious about everything! Music was always a holy and a sacred thing, and suddenly came somebody who had *fun* with music. And he was so overwhelming with this fun, this joy that he communicated, that the whole audience would follow him.

What was special about working with Bernstein? Could you describe, for example, your work with him on Das Lied von der Erde*? (This was recorded with the Israel Philharmonic and René Kollo in 1972.)*
Well, I had done this piece already with Klemperer, with Solti, of course, and with Karajan, so I had worked on the piece with some great conductors. But I really didn't know properly what it was about, what happens in it. I sang it with my musicality, but not with *understanding* it. I didn't realize. Then when Bernstein made the music, he did not have to explain. Somehow I understood it. He took you, I think, to what was *behind* the music. He made you see past the first degree. And so he was with everything. When I sang with him for the first time, for example, the *Missa Solemnis* of Beethoven, it was the same. There were some other conductors who came to me and said: "How can you sing with him? You never know where you are with his beat." Well, I never looked! I heard the music, and I knew from that what he was trying to do. The understanding came through the music.

Bernstein accompanied you at the piano on a number of occasions. (Recordings of Ludwig with Bernstein at the piano include Brahms Lieder *and Mahler's* Des Knaben Wunderhorn.*) How was he in this role?*
He made a whole orchestra. There are some accompanists who say: "Oh, he is terrible, he must always dominate," and so on. But I found him wonderful to work with. He made new all the songs we have known for years and years, and he made a different sound—he came always from the symphony character. He didn't come from the chamber music side. Of course, at the first rehearsal we had problems—we were at the opposite meaning

about everything! It was too slow, it was too loud, it was too fast, it was too soft. Everything I knew, I had to do away with. But then, of course, we met in the middle, and his musicality was so strong that it was impossible not to follow him. His musicality was overwhelming. When I say that it was impossible not to follow, it sounds as if we singers have no character, but I am talking about following the *great* conductors, not the mediocre ones. With the great conductors you will find singers are usually prepared to do what they want.

There are people who have suggested that Bernstein's private life to some extent ate into his energies as an artist. Was this something that you ever felt?

Well, I don't know, because I must tell you, I used to flee from him after concerts! I only went out with him once after a performance, and I knew then that this was impossible for a singer to do, because I never could sing the next day. So I always used to rush away after a performance; I would say, quickly: "Good-bye, good-night!" And he would always say to me that I was so pragmatic and so on. And I would say: "What would you say if I had no voice tomorrow?" So our friendship was close, but only in the musical sense. What he did in life was enough for ten people. And it was good like this. If he hadn't lived his life the way he did, he never would have had this charisma he had. Maybe he ruined his health, but on the other side, my God, he did more than a healthy man! People must live the lives they want to. And it was necessary for him to live this life. I think, you know, he was satisfied with his life. I forget whether it was Harry Kraut or Schuyler Chapin who told me that at the end, when he knew he was ill he said: "It was a great life, wasn't it?" And he spoke already in the past tense. He knew then that he was going to die.

You have spoken about Bernstein's charisma. What was special about the physical element of watching Bernstein make music?

It was not only the watching, but this thing—it is difficult to measure or explain—there are waves that go out from a conductor like Bernstein to the audience. This is the mystery of conductors. They have this special quality, and with it the whole audience becomes full of electricity. That is what a great conductor will transmit.

The other thing with Bernstein that I always noticed—with every performance he was different. And you know the same thing Karajan made also. He said: "If I make always the same tempo, you are in a routine after two or three performances." And so they do it on purpose, to be different. Also it has something to do with their constitution, how they feel when they wake up in the morning! It is the question of how is the weather, how is the pulse; so they are never the same. It is always different from the last performance. And you have to follow. This is I think the reason that conductors mostly take the same singers. For example, with Bernstein, I worked with him all the way from 1968 to 1989.

Some of the orchestral players with whom Bernstein worked have said that he would not always appear secure with an interpretation. Was this ever your experience?

I tell you something, he was always very doubtful. He was never really satisfied with what he did, and he was always looking to do it better. For example, he was very unhappy in Vienna when he was doing excerpts from Wagner operas. Also I remember he said: "What has a decent Jewish boy, what has he to do with Wagner?" There was a love-hate feeling, I think, that he had for Wagner. He wanted to come close, to look deep into the music, but he knew that with Wagner's anti-Semitism it was not really possible for him. I know that Barenboim and Maazel and Levine, they are all conducting Wagner, and perhaps they are really right. But he was disturbed by the character of Wagner. And with other music he was also never satisfied. He would say:

"Was it good? Couldn't we do it better? Perhaps we should rehearse again."

I think this approach is wonderful, because I know other conductors who are very satisfied with themselves. So many of the conductors that you meet, they will say that what they conducted yesterday was a big success, what they conducted today is a big success, even what they conduct tomorrow is a big success! He never, in my experience, talked about his conducting in that way. Also, he treated the orchestra and the singers all as individuals, not as one big unit. In some way he made every orchestra member a soloist. When he worked with the youth orchestra from Sapporo on his last tour to Japan, he played the Schumann Second Symphony [one of the first works that Bernstein had ever heard Mitropoulos conduct], and at one point he was not satisfied with the violins, so he said: "Please, all of you get up when this phrase comes, like you are all soloists." So the whole first violin section got up, also in the concert, and it was very moving.

To turn to the compositions, do you think Bernstein felt that he had failed as a composer?

Ah, but *Mon Dieu*, we are all frustrated in life, we all want to do something else than what we do. *West Side Story* or *Candide*, I think these are wonderful works, but maybe he only thought of them as a light music, operetta music. You know, I think with Bernstein, and with all of us, often the thing that we really can do, it's not so interesting. We always want to sing something else, or compose something else, or whatever. Karajan also felt this. Before he died he said: "I have still so much to do."

Could you describe your final performances with Bernstein, in Candide, *in London?*

This performance from *Candide* in London, I think it was a very sad story. Because we all were ill, and Bernstein was the worst.

We all got this British influenza, but he was already seriously ill. I got the impression that he was just acting like Bernstein. He was not himself anymore. He was sick, and old, and suffering, and he would jump, for example, like the old Bernstein, but he could almost not stay on his feet anymore. You know, this work *Candide* was always the child of sorrow for Bernstein, it always caused him problems. But I think it is *great* music. I love it very much. Also, everything is in it, the meaning of life, the philosophy; for me, this is *the* Bernstein piece.

Do you think it may one day come to be regarded as a more important work than West Side Story*?*

I think so, very possibly. The Overture is wonderful, and the wit of the writing, he really kept it through the piece. He somehow managed to get the wit of Voltaire into the music, the irony of Voltaire.

NOTES

1. Reri Grist played the part of Sophie.

Frederica Von Stade

Frederica Von Stade is one of America's leading mezzo-sopranos, having appeared on the stages of all the world's great opera houses as well as on concert and recital platforms throughout the USA and abroad. Her career began at the top when she received a contract from Sir Rudolph Bing during the Metropolitan Opera auditions. Renowned as a bel canto specialist, her work with Bernstein has included recordings of Haydn's *Theresienmesse,* Mozart's C Minor Mass and a performance at a *New Spirit Inauguration Concert* given for President Carter in 1977. She recently took the part of Claire in Bernstein's *On The Town,* performed (and recorded by Deutsche Grammophon) at the Barbican Hall in London, in 1992. Von Stade began by commenting on questions about *On the Town.*

Was the book for On The Town *somewhat flimsy?*

I don't know what the book was like in its original form; for this performance we have just done bits and pieces, not the complete book. I never have seen the whole show, so it's difficult to judge. I do think that the music is spectacular, so that could be a strong force in it. I think, as admitted in the program, by Betty [Comden] and Adolph [Green], something else that held it together was the choreography by Jerome Robbins. You have to remember—it's a *show,* the plot is an excuse for a show. It's a chance to get in different numbers, of different variety, in that type of a formula, rather than the other way around, with the tune coming out of the plot. Also it was written in 1944, so it was at a time when there was a need to get people out of their doldrums, to "Make 'em Laugh." It was trying to seek that wartime *pizzazz.* That was in there too.

Do you see any problem in opera stars' performing Broadway Shows?
It's fun, although I think it has to be done carefully too. That's something that one has to entrust to the person who is putting it all together, in this case Michael [Tilson Thomas]. What has been fun is the combination. For us [opera stars like Von Stade and Ramey] to get to sing in our own idiom, not just English, but to sing in *American,* is a great treat, though I don't know that on any kind of a permanent basis there is any way we could compete with the Broadway stars! It's a different medium altogether. Also, usually these scores aren't played by symphony orchestras [in this case the London Symphony Orchestra] so the whole thing has a different sound and a different look to it, and that has to be taken into consideration as well. I think Lenny had a great knowledge and understanding and respect for the voice, and a lot of the singing is legitimate, so to speak, and it isn't all Broadway belting. So the mix isn't totally unnatural. It's not as if you were asking Tebaldi to sing it. I have enjoyed doing it, because as a little girl the American musical was one of my loves, and it's part of our heritage.

Did you watch the Young People's Concerts *much as a young girl?*
No. Unfortunately not. But one of the very first symphony concerts I ever heard was Lenny conducting the New York Philharmonic and I adored it. In fact the circumstances of my growing up meant that I didn't have a lot of exposure to classical music, and it wasn't a regular part of my life. So I can't say the *Young People's Concerts* had an influence on me. But I have seen some of them retrospectively, and the great thing about them is that they are so accessible. Lenny took classical music off the shelf, and put it out there, made it accessible to a wide American public. It wasn't just the knowledge and the intelligence and the

brilliance—it was the way Lenny did it. Lenny loved the audiences of those concerts. If I had to give my one impression of Lenny, it would be to say that he loved people. He loved being around them, loved the challenge of them.

Could you talk about your work with Bernstein as conductor?
I remember my audition for Bernstein; I think I sang the little song at the end of the Mahler Fourth Symphony. I remember sitting next to him at the piano, and he was helping me to see what was going on—he was teaching me, really—he loved teaching people. And it was never in an authoritarian way, it was never "You should be doing this," but rather "Do you understand this?" or "This is what's happening there," or "Try this," and like a lot of great creative people there was an energy about him that was just startling. Another shining point about Lenny was his rhythm. Whatever tempo he decided to take you never lost a sense of the continuation, and if he had chosen a slow tempo, he would make it work so that you didn't notice its being slow.

It was a great thrill to work with him. I did the *Harmoniemesse* of Haydn [1973], and then we did President Carter's Inaugural Concert and I sang "Take Care of This House" which is from one of his shows [*1600 Pennsylvania Avenue*]. He was always enormously affectionate—for example, in Salzburg, if he happened to attend a concert you were in, he would always come round to the dressing room afterwards. I sang at his Seventieth Birthday Concert and I remember his managing to juggle the public and the private elements of that exceedingly well. The last time I worked with him was in 1990, on the recording of the Mozart C Minor Mass, in Germany, and I don't think he was feeling particularly well, but he didn't show it. I don't know whether he knew how sick he was at that point, but he certainly didn't let on to us. It was very moving to watch Lenny, in this beautiful church, going through the Mozart C Minor Mass with such great reverence and humanity. That was the last time I saw him.

Do you think there was a conflict for Bernstein between composer and conductor?

I don't know. I would guess that along with the dissatisfaction that any performer has, with always going for more, going for better, going for richer, that a composer would be even more tortured. Also Lenny was such a public person, and public life takes an awful lot of energy. I think some of the energy that he needed for composition was absorbed by the demand on him publicly. Like a lot of performers he was everything—a conductor, a pianist, a teacher, a composer. I imagine it would have been very hard to keep it all going, to keep the ball in the air, so to speak. I suppose that his predicament is to some extent portrayed in the song in *On the Town*—"We'll Catch Up Some Other Time." I think that was possibly his story. Also he was such an accessible person, that it would have been very hard for him to turn down people. I think one man can only do so much.

What do you think will last of the compositions?

Of course *West Side Story*, and I think *Candide* and maybe *On the Town*. I think these works are being presented wonderfully well now, and carefully, and I think that is very important for the future. I rented the movie of *West Side Story* the other day for my kids, and even though it looks a bit dated, my kids just sat there, rapt. I've never gotten sick of *one note* in *West Side Story*.

Do you think he was more comfortable in the realm of the musical theatre, rather than serious composition?

I suppose, yes, he was such a Showbiz guy.

Orchestras VII

Members of the New York, Vienna, and Israel Philharmonics

The New York, Vienna, and Israel Philharmonics are all orchestras with whom Bernstein had a special affinity, in spite of a schedule that included several other great ensembles. The voices recorded here include: Stanley Drucker (Principal Clarinet, New York Philharmonic), Jon Deak (Associate-Principal Double Bass, New York Philharmonic), Rodney Friend (Ex-Concertmaster, New York Philharmonic), Rainer Küchl (Concertmaster, Vienna Philharmonic), Yaacov Mishori (Principal Horn, Israel Philharmonic), and Avi Shoshani (Secretary General, Israel Philharmonic.)

The NYPO, the VPO and the IPO were probably the three orchestras closest to Bernstein. What was it like to play under his baton?

STANLEY DRUCKER I first joined the New York Philharmonic in the 1948–49 season at the age of nineteen. And over the years I would say that the period under Bernstein was a very special one. It was the essence of *live* performance with him. And we did a lot of new repertoire, a lot of American composers—I remember wonderful performances of Ives, for example.

RODNEY FRIEND I recall the first time I played under Lenny with the Philharmonic. We (the orchestra) were sitting on the stage waiting for him to turn up for the rehearsal. It was like the beginning of term at school. We waited and waited, and then he arrived and the feeling was one of great warmth and comradeship. Lenny had grown up with a lot of the musicians, and it was as if he were being welcomed back into the family. And when Lenny embraced half of the orchestra it was not, as some people have suggested, an affectation on his part. Neither was his behavior on the podium. I remember when I saw him in the sixties conducting in the Festival Hall in London, I had thought that he was rather showy. But then when I worked with him at the New York Philharmonic, I realized that it was not a show, but a natural expression of what he was feeling. Everything about Lenny from the way he walked onstage to the way he spoke to the players was genuine. And another remarkable thing that one felt while working with him was that there was no fear. There was no anxiety that, as with some conductors, discipline of the wrong type would come between the musicians and the music. Working with Lenny was all to do with color and characterization and the composer. It was never about personalities.

JON DEAK Playing under Lenny was like being there at the moment of the creation of a new work. It was as if, and I experienced this with him many times, he was writing the work himself. That goes for anything from Haydn right through to Mahler and Stravinsky. Also he demanded total commitment from the music. With him it was a matter of literally grabbing the work by the throat. He wasn't satisfied with a performance, if, for example, we were playing the *Pathétique* of Tschaikovsky, unless half the orchestra members and half the audience were crying along with him. Playing with Lenny was always a frightening but exhilarating experience. I remember the Tchaikovsky as a striking example of his coming at something totally fresh. He would say: "Wow! You know, it just came to me what that phrase meant!" And your heart would go out to him. And when you think how well Bernstein knew this symphony—I mean, I know it memorized, and he would have known it ten times better than I do—it was extraordinary.

RAINER KÜCHL I was with the Vienna Philharmonic for Bernstein's *Rosenkavalier* production, and it was the first *Rosen-*

kavalier I played in my life and very exciting to play with him. I have played this opera many times since then with other conductors, and with Bernstein it was not the typical Viennese way. But he had a feeling for the Viennese waltz rhythms—these he played very finely. He was a conductor who made all his work with the heart. So he touched everybody, the orchestra and the audience.

How easy was it to follow the choreographic Bernstein podium manner?

JON DEAK It *was* sometimes difficult to work out what he was doing. When Lenny started the Beethoven Fifth, he would suddenly shake his fist and stamp on the floor. Now *where* do you put that first eighth note? We would follow the leader of the first violins. Another example, I remember at the end of *L'Après-midi d'un faune* there are these little *pizzicati* for the bass section in the orchestra, and where do you put them, how do you place them with these little curlicues that Lenny made in the air? What Lenny assumed, I think, was a certain basic knowledge of a work, and an inner rhythmic drive and orchestral cohesion, which meant following the leader of your section and also your partner. So his interpretation was much more focused on the broad structure, and the shape of the phrase, and the emotion on any given note, rather than conducting four square.

STANLEY DRUCKER If you watched Mitropoulos you sometimes couldn't figure out what he was doing [technically], and Lenny was the same way. In as much as he did do complicated music it was amazing. But both conductors could get results. With Mitropoulos it was always very exciting; just what he could draw out of the orchestra in terms of dynamic range was amazing. And Lenny's results with an orchestra were very similar. Every concert became an event, and you went away feeling clean, feeling exhilarated.

RODNEY FRIEND In the sixties I worked a little with George Szell, and Szell's orchestral technique was probably the greatest I have ever come across. Lenny did strive for that, he wanted that, but he was perhaps more interested in the emotional dimension of the music. And when you worked with Lenny, there is no question that he gave one hundred percent all the time, and he took you there with him. He inspired the players. An orchestra like the New York Philharmonic doesn't really need a time beater. They don't need somebody to wave a stick at them. What orchestras of that level need is a great personality, who will bring them together as one unit and with one idea. And Lenny did that. I remember a tour we did to the Far East with Lenny and we took Mahler One. Imagine, playing every few nights a piece the size of Mahler One! But with Lenny every night became more wonderful. Every night would be different, with a new vision or a new feeling.

YAACOV MISHORI In 1982 we [the Israel Philharmonic] did a tour with Lenny to Mexico and the United States and one of the concerts we played was in Houston. We played *Francesca da Rimini.* Five minutes before the end, where Tchaikovsky is describing Francesca da Rimini and her lover burning in hell, Lenny did a huge jump. And suddenly we couldn't see Lenny. No more Lenny! We didn't know what had happened. We thought maybe he had fallen into the audience, and then we heard him calling: "Go on, go on!" He even called to us in Hebrew, I remember. He had fallen in front of the celli and they helped him back to the podium while the orchestra kept playing! It was a frightening experience, but fortunately Lenny was not really injured.

RAINER KÜCHL Bernstein would get very excited in a performance. Maybe at first I think people were a little surprised by all the jumping. Also, sometimes in a big excitement, he would make like a boxer or he would open out his arms and he would knock our music over! The desk was very close to him sometimes, because the stage is not so big at the *Musikverein.*

After Bernstein left the New York Philharmonic, in 1969, he was asked more than once whether he would return to the orchestra. (When Zubin Mehta arranged to leave the NYPO, Bernstein was the first conductor the management contacted.) Was this something the players were aware of?

JON DEAK I think Lenny was constantly, if informally, being approached with raised eyebrows about the possibility of his coming back. There was one occasion when he turned around to the audience, totally compulsively in a concert, I think in the late seventies, and said: "This is *my* orchestra. I just wanted you all to know that. And I'm coming back!" I think he was swept away emotionally by the performance of that evening. Anyhow we [the players] went to the management and they didn't know any more than we did. Sometimes also I remember Lenny would come for a stint of guest conducting and wrinkle his eyebrows in horror and say: "My God, what have they been doing to my orchestra? Is this how you play this piece now?"

RODNEY FRIEND During the Boulez period, and then after that when Zubin Mehta took over, one still somehow thought of the [New York] Philharmonic as Lenny's. Of course, Boulez is a unique musician and he is an amazingly advanced musical mind. But perhaps he wasn't entirely comfortable with the Philharmonic at that particular time. Then with the arrival of Zubin Mehta the problem was that he was a conductor of the traditional school, whose repertoire was similar in many areas to Lenny's, and it was difficult for him to live with the orchestra for so long under the shadow of Leonard Bernstein.

STANLEY DRUCKER I wish Lenny had stayed on longer with the Philharmonic [after 1969]. But orchestras with histories as long as the Philharmonic have an ongoing situation. You can think of certain provincial places where somebody might stay for thirty or forty years, but in New York, which is probably the toughest place of all, that is unlikely. Don't forget the Boulez period was very fine too. We made some very wonderful recordings under him. People said he was cold and so forth, but he could take the most complicated score and prepare it quickly and with total confidence and reliability. His balancing and his ear were marvelous—I thought that was an exceptional time, also an innovative time.

What was Bernstein like as an accompanist on the podium? Vladimir Horowitz reputedly told Bernstein: "You can't accompany. The more important the player the more you steal the show."

JON DEAK This was an area that was always difficult for Lenny. He would throw himself into a piece so much that he would want the soloist to go along with him. I think he had trouble following somebody else's independent idea, which is not actually uncommon with certain conductors. I am not sure I understand the particular psychology of this, because as a player (even as a composer) I love to work *with* other performers. It is true that Lenny managed to collaborate in the world of the musical theatre, so I think that philosophically he could collaborate with other people, but as a performer, once the kinetic thing started he wanted to be in total control. So it is interesting that a lot of the programs that we did with him later on were without soloist; decreasingly as he got older would he use a soloist.

STANLEY DRUCKER I remember very clearly the Brahms First Piano Concerto with Glenn Gould. It's a famous story, Lenny turning to the audience before the piece was to start—he was probably a little overwrought—and saying he would not take responsibility for the piece or the tempo. They obviously couldn't get together on it, their minds did not meet. And Lenny told the audience that straight out.

As an interpreter, many people found Bernstein's music-making, particularly towards the end of his life, too personal and egocentric. Was there this perception among the players?

RODNEY FRIEND Well, the *Nimrod* Variation from *Enigma* [recorded with the BBC Symphony Orchestra in the 1980s] was very slow and I actually asked Lenny whether he was being serious, at the rehearsal. And I remember it was being filmed for television and he said: "You bet your last cent I'm being serious!" I had played the Elgar many, many times with [Adrian] Boult and discussed it with him, also recorded it at least twice. Now Lenny's version *was* very drawn-out, and after the long sustained G that leads into *Nimrod* it became about half speed. It was as though he was searching for something, and whether indulgent or not, whether tasteful or not, it was honestly felt. However much he seemed to be milking the piece, it was genuine.

AVI SHOSHANI Toward the end in say the Brahms symphonies he became very slow. It became very long, but it was still coherent and convincing. He was not doing that because of wanting to be different or to be special. The tempo that he took was what he felt. And you never thought he was dragging the music, even though at the end of a concert you would see that a symphony had taken maybe seven, maybe even ten minutes longer than usual! For instance, one of the last works that he recorded with us was the Dvorák Cello Concerto with Mischa Maisky. And I remember Lenny said: "Why are you treating this work as a concerto? I think it is a symphony." He wanted a grand conception. And if you listen to the recording you will see that it works as a symphony, it is convincing. He never imposed upon the music, it came out very naturally.

STANLEY DRUCKER When I played the Copland Clarinet Concerto for the last time with him—these were his last recordings at the Philharmonic, in 1989—he wanted to do the opening—a slow-moving melody—almost like a Mahler slow movement. It took a little getting used to. It was like a prayer. It was a question of understanding what he was doing, and then from the cadenza on through the rest of the piece it was more or less what I hoped for, and there was a tremendously exciting, driving conclusion.

How does one define Bernstein's relationship with his orchestras? He seems to have engendered a camaraderie among the players . . .

RAINER KÜCHL I remember he came on stage and we made music, and all the problems we took together. It was not: "I am the master and you are the slave." He treated everybody equally, and he was a strong personality, so everybody followed him. The orchestra [the Vienna Philharmonic] also did a lot of work with Karajan. And it is fascinating what different people Bernstein and Karajan were. For us they were like North Pole and South Pole! With Karajan, you could say at the beginning he was a little like a general—you can see in some of the old films he is like this—but not in his last years.

JON DEAK As a player, I can say that I don't know why there should ever be a reason for a conductor to browbeat you. We need the suggestions of a conductor, and the discipline so that if it's not going well we will keep going over it, but that's as far as it goes. With [Kurt] Masur today, we have that wonderful quality—we don't give up until we have got it right. Bernstein's discipline existed purely on the passion of his artistry, and of course not many people can pull that off.

YAACOV MISHORI I can tell you as someone who played under his baton for many, many years—Lenny was a democrat. He *trusted* the musicians. When Lenny said: "What about an accent here and a diminuendo there," he would always ask, and often invite your opinion also. He would never demand things. I remember once we were playing *Das Lied von der Erde* and I went to him at the intermission and said: "Can you do me a favor and give the horn section an entrance two or three bars before Letter D?" He told me: "My dear Mishori, I don't conduct bars, I conduct phrases. I trust you, I trust you to come in there." There are some conductors whom I really feel tension with, when I play with them. With Lenny I always felt relaxed; he let me play, and interpret my solos as I felt them.

Many colleagues, Christa Ludwig among them, have commented on a quality of self-doubt in Bernstein's work. Was there ever this perception among the players?

JON DEAK Perhaps because people were always taking issue with what Lenny did, he himself was always extremely self-critical. In fact it seemed to be almost a problem with him, something that he really suffered from. He needed to be reassured constantly as to his artistry. I think he sometimes felt a lack of confidence in his own direction—particularly as a composer—in terms of reaching the people. Lenny believed that there ought never to be a gap between the creation of a work and its acceptance by the public. There was something wrong if that happened. He wanted everything he did to be understood.

YAACOV MISHORI In September 1985 we played a tour with Bernstein in Japan and the United States, with Mahler Nine. According to Bernstein himself, the Mahler Nine in Tokyo was one of his best performances, if not *the* best, of the symphony. He said it was a pity we did not record that performance. After he finished the work he stood in the wings, crying like a baby, without coming back to take his bows. And you know what was very interesting about Bernstein and this piece—he would sometimes say to me: "I really don't know whether this is the way to perform it or not." And yet he did it so wonderfully. We would do it six or seven times, and every time it was different, he would try to find new meanings in it.

Bernstein brought the Vienna Philharmonic on tour to Israel in the 1980s. Was this a controversial move, in the light of the election of Kurt Waldheim and Vienna's anti-Semitic history?

AVI SHOSHANI When Lenny brought the Vienna Philharmonic on their first visit to Israel it was a huge success—it was judged or looked at only from an artistic point of view. It was always mentioned in the press and so on that it was a historical moment, but what really mattered and what made people happy was the quality of the concerts. Lenny had a close relationship with Vienna, also with Munich, which was not so easy for us in Israel, but one cannot judge a person like Lenny in the way of a normal human being. With him there are other things involved—creativity, artistry, music-making which make the criteria different. Let's put it this way: I didn't like it, but I didn't have a problem with it.

YAACOV MISHORI There was some tension when Lenny brought the Vienna Philharmonic. I think it was mainly because of Kurt Waldheim. But the music lovers in Israel—we have more than thirty thousand subscribers—try not to mix music and politics. And the Vienna Philharmonic was there to make music. Also, of course, the public loved Lenny. Some of my colleagues in the VPO horn section told me that they were surprised; they had expected possible demonstrations, but there was nothing like that. A most moving experience for me was an Israel Philharmonic concert with Lenny in Germany in 1978. We played some of his compositions, works like *Chichester Psalms* and *Kaddish,* a mere five hundred meters from the Reichstag. We were playing a short distance from where the orders were given for the extermination of the Jewish people! Those concerts were like a revenge for us, on Germany. They had tried to exterminate us, and here was a Jewish orchestra, with a Jewish conductor and composer, playing in this place, and playing compositions based on Jewish tradition.

Although many of Bernstein's later recordings were done live, he spent a good deal of time in the recording studio. What was he like in this environment?

AVI SHOSHANI He was *extremely* human. Sometimes Lenny would be nervous, or tired, or he would be swearing, but one would understand. Most of the Israel Philharmonic's recordings for Deutsche Grammophon were with Lenny. (With Zubin [Mehta], for example, we recorded for Decca and EMI and for a while CBS.) But recording with a German company was not really a problem. It happened at a stage when one could look at it purely from a professional point of view. I remember our recordings of the three [Bernstein] symphonies. Of course, sessions are always full of tension and nobody is ever satisfied. There is always the feeling that there isn't enough time. The knowledge that something is going to be put on a tape forever, pushes everybody to the edge of their nerves and one can never really relax. But with *Kaddish* or *Jeremiah* or *Age of Anxiety*, we were ready to sit for as many hours as Lenny needed in order to accomplish it. And he would go overtime or he would be a little bit late for rehearsals,[1] but we were so much in love with him that we accepted it.

STANLEY DRUCKER During Lenny's years as Director of the New York Philharmonic we used to record very, very often. Also of course we then had a fifty-two-week season, as opposed to twenty-eight weeks when I first joined. In those days recording sessions would go on and on with seemingly no limits. I remember when we were doing, for instance, a Mahler Symphony, after recording for many hours, Lenny would say: "Okay, now let's perform it!" He became totally immersed in the music, and he pushed himself—and his players—to the limit.

How does one place Bernstein as a conductor? Where does he belong in this century?

AVI SHOSHANI For me there was Leonard Bernstein and then the rest of them. I can only judge from the big conductors that I have seen conducting. If one talks about Zubin [Mehta], Daniel [Barenboim], Claudio [Abbado], [Riccardo] Muti, I am speaking about that category—it's Leonard Bernstein and then everybody else. And I think most of them do realize that. It's like we used to say: there is Jascha Heifetz and then there are all the other violinists. Also with Lenny he was not only a conductor —he was much more; he was a composer, a teacher, and all of that created a certain quality that made him different. I have never experienced a concert with Karajan. I must admit that although I met him personally I have never seen him conduct, and I don't think it is fair to make judgments from a video. But, you see, I think Karajan was lacking in certain qualities that Lenny had—the warmth, for example, the humanity. I think if you had looked at the dressing room of Lenny after a concert and the dressing room of Karajan after a concert that would have told you everything!

RAINER KÜCHL He was the greatest Mahler conductor of his time, I think. I don't know how he compared with somebody like Bruno Walter, but I think in the period from the sixties, he was the greatest Mahler conductor. Just after I came to the orchestra [the Vienna Philharmonic] we started the Mahler cycle with him, for record and for video.

YAACOV MISHORI Lenny was different from other conductors. As you know, a conductor is like an architect with the music—he is responsible for the dynamics, for the balance of the instruments, for the ensemble, and so on. Lenny was much, much more than that. As a musician, I feel he almost hypnotized the players. He was so spontaneous, and he inspired the orchestra so much, that we felt that we had to give everything for him. While I can appreciate playing under Maestro X or Maestro Y, with Bernstein I felt I was a part of the music.

NOTES

1. Avi Shoshani has added the following: "Toward the end Lenny had a problem that he was awake most of the night and sleeping during the day—his timetable was turned around—and this made morning rehearsals very difficult. As much as we [the orchestra] were ready to put a rehearsal at 11:00 instead of 10:00 we couldn't make it later than that. He had for quite some time this problem that he was at his best in the late hours of the night—or early hours of the morning. But as I told you, the musicians here in Israel were ready to do anything for Lenny, and so they sat and waited for him. The orchestra was ready to do things for him that it would not do for any other conductor. To be prepared to put in longer hours is very unusual for musicians nowadays. For him they would do it, they would do anything he asked them."

VIII
Bernstein on Broadway

Carol Lawrence on West Side Story

Carol Lawrence began her career as a soloist with the Chicago Opera ballet. She made her Broadway debut in *New Faces* of 1952 and also appeared in the film version. Other productions in which she has been seen include *Me and Juliet, Plain and Fancy, Shangri-La,* and the City Center revival of *South Pacific.* She was also seen in *Ziegfeld Follies.* At the age of nineteen she met Leonard Bernstein for the first time when she was chosen for the part of Maria in the opening production of *West Side Story.* In the historic first production she played opposite Larry Kert's Tony in the preliminary "try-outs" in Washington and Philadelphia, and then at the triumphant first run at the Winter Garden Theater in New York. The interview began with her description of the events leading up to that first production . . .

Could you describe your audition for the part of Maria before the show opened in 1957?

First of all you have to know that Jerry Robbins had the concept of *West Side Story* some ten years before the curtain actually went up on Broadway. Now, Robbins and the other collaborators [Bernstein, Sondheim, and Laurents] had been involved in auditioning people all over, and I was one of the last to audition for Cheryl Crawford. If you remember, she began as producer of the show. I was signed with William Morris at the time, and was asked to come, along with a laundry list of people, to her office, to audition for one of the principal roles, the part of Maria, which was what I was being considered for.

Cheryl Crawford's office was very small. It was like a little rectangle, with a desk at the end of it—I'll never forget, she sat behind it—and then on a settee, very near the door on the perpendicular wall to her desk, were Jerry Robbins, Arthur Laurents, Steven Sondheim, and Leonard Bernstein!

The "Big Four"?

The "Big Four," and as close to you as your nose. I could have touched them if I extended my arm. Which I didn't! I just stood there, frozen. Then Peter Howard, at that time my rehearsal pianist—he was a wonderful accompanist—went into the intro of "When Does This Feeling Go Away?," my chosen song. It was a lovely little ballad, and I felt very secure in it. Lenny said: "Thank you very much; now would you sing an aria?" I thought: Terrific, I've made it to the next plateau! At which point my agent, Bruce Savan, who was in the room keeping score, put his hands over his ears and knelt in the corner like a little mouse. I was aghast. Here was my agent, whom I was looking at, pantomiming: "Don't, for God's sake, *don't* sing an aria." Peter looked at him and went right into the introduction of what I had prepared. I sang and when I finished, Lenny said: "That was wonderful, Miss Lawrence, I would like to hear you in the theatre."

You see, Bruce [Savan] miscalculated. He didn't think my voice was right for the part. That was his judgment, and thank God he was not making the decisions. My voice was that of a teenager—I was nineteen at the time—and that's what they were looking for. They were *not* looking for an operatic sound.

Sondheim especially seems to have been very specific about not wanting a "trained voice."

Oh, but it was Lenny's choice too. I certainly didn't pretend to have an operatic sound; I had started from early on as a dancer. I had been a scholarship student at Edna McCrae in Chicago, where my training was primarily dance, and I had my own night-club act when I was thirteen. I was being groomed for Broadway, not for opera. Still, singing was a part of my Italian heritage. I had sung in the choir at church, and led the sextet at high school and so on. But it wasn't a trained voice. And they were searching, in truth, for that sound, for something very youthful—something unschooled, but pure.

So then I was called back, to come into the theatre, but in the interim Cheryl Crawford dropped the show.

Was this because she thought it would be a commercial flop?
Yes, she thought it would be critically acclaimed and that the American public would never buy it—teenagers in blue jeans, not a star in sight, a tragic ending, trying to play *Romeo and Juliet* in the ghettos of New York, with Lenny's operatic score and more dancing than anybody had ever heard of—she said: "Look fellas, it will never fly." She didn't believe in it, but then she was in the majority of opinion at the time. The problem was that all the know-it-alls in theatre didn't take into consideration that we were touching a nerve in the American psyche, and that the music of Leonard Bernstein was so magnificent that you could not hear it for the first time and not be moved.

Anyhow, after that I never saw Cheryl Crawford again, after my first audition. And there were another twelve auditions to come—I did thirteen overall! And it was not a big number. You have to understand that Jerry Robbins was the motivating force in all of this. He was the eternal perfectionist. The fact that one can never attain perfection did not deter him for a second. That was what he wanted and if he ended up killing you in the interim, well that was okay too!

Apparently he had the cast in tears for much of the time . . .
Oh, in tears, and in fears, and in trauma, and in shock—he had them every which way. You have to remember that this was thirty some years ago, when that kind of treatment was allowed. And you have to realize that Jerry came from a ballet back-

ground in which the choreographer is the master and the *corps de ballet* the absolute slaves. Dancers get used to that treatment only because it works. When you intimidate and humiliate a dancer and say: "You *can't* jump higher, you *can't* jump further . . . " his or her attitude is: "Goddamn you, I'll show you." And you do it, because the adrenalin flies through your system, and you literally do it to show them up. And so it's rebellion that the choreographer is calling upon to serve his ends. Now, that works in a dancer, but it does not work in an actor. You cannot force, for example, an actor to be more poignant. But you see Jerry had never directed before; he had come from where he was—as a choreographer—with the only tools that he knew how to use. That said, I am not faulting his abilities—he's a genius of a choreographer, and I would submit myself to that brutal treatment tomorrow if there were a show that he was doing that I was right for. So this is not a put-down; these are merely the facts. And in that arena—thank God—was Leonard Bernstein, because Lenny was the opposite end of the pendulum sweep. His role was as the gentle teacher, the logical, compassionate, caring, and articulate teacher, who inspired you so that you wanted to please him more than life itself.

Would Bernstein work on the score with you?

Oh, he was my and Larry's rehearsal pianist for the show! We would go into a room with a piano and work on one of the numbers, and he would literally bind our wounds. He would heal our psyches and prop up our self-esteem so that we had the courage to walk back on stage and try again. I felt that just to walk into a room where Lenny was, was to know that you were in—I think—the presence of genius. And you know genius can take on many, many qualities of egocentric behavior that are very unappealing; that never happened with Lenny. He was a person who when he spoke to you or looked into your eyes would give the impression that he didn't care about another person in the world.

Many people have spoken about that particular quality.
Oh, it's absolutely true. John F. Kennedy was the same way. There are those charismatic people who own that quality, and it's absolutely identifiable. It's like when you break a bone, it's a pain that you will always remember. It's really that clear, that definite. He would work with us by the hour, and he would say: "If it doesn't feel good in your mouth just tell me and I'll change it." And we would say: "Just tell you *what*? No, we'll do whatever you want!"

He always seems to have been ready to change things—also with other collaborative efforts.
Yes. That to me is also a sign of his genius, because you knew that he was capable of coming up with something that was even better. Most people don't think they can; he *knew* he could. And you heard him do it, repeatedly. And he never lost his temper; during all that time he never said a cruel word to anybody. He was the epitome of patience, and also endlessly resourceful—he had to be. I remember one morning when we were rehearsing in Washington, Jerry had asked Lenny to change something in the *Scherzo*, and to bring in some new music. We were all sitting on the floor and Lenny brought out the score and played it for us and it was beautiful. And Jerry turned and said: "That's worse than what you had before. Go write it again." This was in front of the company.

And what did Bernstein have to say to that?
He picked up the music and went and wrote it again. He didn't even say: "Jerry, do you mind telling me that alone?" He didn't even ask him to give him the respect of being the composer of that incredible score.

So did Bernstein always defer to Robbins?
He always did. You know they loved each other, and their roots went back to *Fancy Free*. And that's the way Jerry is. If you knew

Jerry you knew that that was the only way he knew how to say it. I don't think he necessarily meant to be that cruel, and maybe he didn't even view it in that light. Jerry is a strange human being, and Lenny adored him and forgave him and knew that ultimately it would be for the benefit of everyone concerned. And it was, but I kept saying: "Couldn't he learn along the way to say 'please' or 'thank you' once?"

But going back to where we were before—once Cheryl Crawford left they didn't have a producer, and Steve [Sondheim] called Hal Prince, who was just opening a show [*New Girl in Town*] in Boston with Bobby Griffith, another theatrical giant, and they came to New York, listened to the score, and decided to do it. [Hal Prince, an old friend of Sondheim's, had initially turned *West Side Story* down.] And so they (and Lenny) became the balance in the whole thing, they kept Jerry from killing everybody! Hal and Bobby took over the production and I began auditioning for them. Jerry wanted to see all of the components juxtaposed—he wanted to see the chemical reaction of this pairing and that pairing of Maria and Tony, he wanted to see what would happen if, for example, Bernardo was Riff, and so on, so they kept calling me back.

Apparently you and Larry Kert were introduced by Ruth Mitchell (the stage manager), and then the "Big Four" asked you to walk onstage together.

Yes. You see Larry had already auditioned for the parts of Riff and Bernardo, and he was not right for either one. Then Steve saw him—Steve and Lenny were actually integral in the choosing of both Larry and me—do an industrial show in which he sang this ridiculous calypso number "No More Mambo." It was very, very high ranged, and when they had auditioned Larry he hadn't sung that way, because neither Bernardo nor Riff needed to have range. But Tony did. Lenny had said that they were looking for a blond, Polish-looking person to play Tony, and

Larry was dark and Jewish. Steve said to him: "Why haven't you auditioned for Tony?"

Larry said: "Because I'm not tall and blond and Polish!" And so Steve told him he would have to come in again, with the idea of being considered for Tony. He passed that audition and at that stage I had passed through some twelve auditions so that I was being considered in the last running for Maria. I always wore the same little pink dress, with my hair exactly the same, and at the end of every audition I would say to Jerry Robbins: "And I also dance, Mr. Robbins."

He would say: "Uhuh. Next!" He would never even consider looking at me as a dancer. I thought I would have an edge for this wonderful choreographer if Maria could dance, but it never occurred to him. Now Lenny and the others all said that they wanted to see us together, so Larry and I walked onstage and then they asked us to come back. They would always hand you the scene that you were going to read, and then take the script out of your hands, lest you show it to anybody. For some ungodly reason I had the guts to say: "Excuse me, Mr. Robbins, since you know that you are going to have Larry and me reading this scene, would you allow us to take it home and memorize it? It's so hard to do the balcony scene—the love scene of all time—with a piece of paper in your hand between you." Now, you see, I was taking some of the control away from him, and he was dumbfounded and very begrudgingly said: "Oh, okay, you can have it." So they cut out those two or three pages or whatever and gave it to Larry and me.

Now, did this include the song "Tonight"?

Yes, that was "Tonight" and the balcony scene. So Larry and I worked once together—dear Larry told this story better than me—he said: "I was so scared, I didn't know what I was doing. She just handed me the script and said: 'MEMORIZE IT!' He was very shy actually. And you can imagine how badly I wanted the part by this time—just reading and hearing it—and by then I had heard the music. Everybody in the world wanted it.

So, Larry and I worked together and arrived on the day at the theatre to do the balcony scene. Jerry took me aside and told me to hide somewhere on the stage—without telling Larry. His plan was that once Larry had sung "Maria" he should try to find me, and we would then do the balcony scene. So now all the confidence I had from knowing the words was taken away. And that was what Jerry was constantly judging—whether you could be thrown from a tenth floor window and land on your feet. That's what he always wanted, because he wanted the show to be constantly spontaneous and real.

I looked around this empty, bare stage and I saw on the back wall this little steel ladder that led to a steel grating—it was a fire escape. I climbed it in my little pink dress and knelt down in the shadows of the wall. They called Larry in and Jerry said: "Okay, I want you to sing 'Maria' and when you have finished 'Maria' find her and do the balcony scene." Now, poor Larry, what a line to throw to anybody. So he sang "Maria" and he was looking right and looking left and he couldn't find me because I was directly behind his head. Who's going to turn around when you're singing in front of Leonard Bernstein and the people you're trying to impress in the audience? You can't turn your back. So there was dead silence in the entire theatre when he finished "Maria," and he couldn't find me, so I cheated—I whispered: "Tony!" He turned around, saw me perched there, and in two bounds he was on that grating. I don't know how he got there, except that Larry was a stuntman, you know, so he was tremendously athletic. And when Jerry saw that kind of spirit he was *very* impressed. So we did the balcony scene. And then Larry leapt off and we sang; "Goodnight, Goodnight . . . " And as he walked off the stage they applauded, which was the first time that they had ever done that. Then they went into a little conference. And later they told us that we were their Tony and Maria.

From then on it was a matter of literally *becoming* those characters. Jerry wanted everything so thoroughly and so quickly that every fiber of your being had to be at his command. You had to justify every single word that you said, *with* a subtext and a basis in your own life. He was Stanislavsky reincarnated. [Stanislavsky's "Method" advocated an intense identification between actor and character.] He never called us anything but our character names when we were rehearsing. Once we came to rehearsal we were never allowed to even communicate with the opposition. It was a battleground. The right side of the stage was earmarked for the Sharks and the left side for the Jets, and you didn't cross that line. He would purposely incite people to antagonistic behavior. And he was brutal, he would humiliate us, always in front of the entire company. It was never in the privacy of your dressing room. Instead of saying: "You're just not warm enough in this scene," or "I don't believe you here," he would say: "You are the most talentless idiot I've ever met in my life, why can't you GET this?" It was like being cut in two, and that would be the moment when Lenny would come to Larry and me and say: "Let's go and do the balcony scene," or "Let's work on 'One Hand, One Heart'."

Much of Bernstein's music in West Side Story *seems to be difficult both in terms of range and rhythmic complexity, particularly for "untrained" voices. How did you find it?*

Oh, it's an operatic score. It's almost a three-octave range and *very* difficult rhythmically—you're never in one meter for more than sixteen bars! It was just taken for granted that we would be capable of doing it, and with Lenny we had a genius teaching us. It came from the horse's mouth, and he was an *articulate* horse! Leonard Bernstein was without doubt one of a kind, and we were blessed to have him, because *West Side Story* would not I think have had a prayer otherwise. We had to be in the hands of somebody who was capable of replenishing whatever it was that Jerry wanted—the palette changed daily. And Lenny was always capable of giving us a new color that wasn't there before. It was a joy to be in his presence every day.

You have said that you think West Side Story *is an operatic score. What do you think of opera stars performing the work, as in Bernstein's 1985 recording for DG?*

Well, you have to remember you're talking to a Broadway performer here, to an actress, and I feel that first of all in *West Side Story* you are dealing—in terms of the character of Maria—with a sixteen-year-old girl, who has just come from Puerto Rico, and who is in love for the first time—she is not going to have pear-shaped tones coming out of her mouth. That is against every truthful portrayal of the character, and every fabric of the play. If you want to hear the score sung perfectly, then do it, have your opera stars, because it is operatic in range. But then don't call it *West Side Story*, call it *Arias from West Side Story*, or the operatic version. In the operatic recording there was also miscasting—Tony was being played by [Jose] Carreras, a Spaniard, when he was supposed to be an American, and that defied the truth of the character. I saw the television film of that, and you could see the pain on Lenny's face when he kept saying: "No!" I've never seen him so angry.

Sondheim had always been specific about needing "singing actors" for West Side Story; *do you think he was clearer about this requirement than Bernstein?*

When he was working with us Lenny never wanted it sung like an opera. He would always say: "Sing it from your heart. I don't *want* pear-shaped tones." Don't forget, *West Side Story* takes place on the streets of New York and we were schooled by him to make it *conversational*, and that's why it worked. With musical comedy, and this is one of the things that makes it so hard to convey to kids, the singing should come out of an emotion too big to be revealed in words. It should not sound like you're suddenly Tebaldi, or whoever.

Now, to return to the world premiere in Washington, there were apparently problems with the choreography because of the size of the stage . . .

Yes, the stage was too small, it was unbelievably cramped, and we had to spend from morning 'til night redefining the ballet. [*West Side Story* had been choreographed for New York's Winter Garden Theater.] Jerry was frustrated and angry. We lost more than six feet of the front apron that we needed.[1]

So, Jerry spent the whole time re-staging "The Rumble" and the "Ballet" and the "Prologue"—there was so much dance, almost nothing but dance in the show. Now, "I Feel Pretty" was Jerry's least favorite number in the show, because it dealt with the inane behavior of a young woman in love—it was something he didn't understand and didn't care about and didn't feel was integral to the work. He actually wanted it cut. It was the only time in the whole show that I smiled. The rest of the time I was saying: "Please be quiet, my parents will hear you," or "SHHH!" or whatever. Whereas in "Pretty" I said: "I feel wonderful because I'm in love!" And of course two seconds later Chino comes in and tells me Tony has killed my brother and from then on I'm running. Anyway I loved the number—apart from Lenny's beautiful music, it was a fun, wonderful moment which was very necessary in the show, because we had so few of those. And at the end of the day [of final rehearsal] I realized that we had not rearranged all of the choreography for "Pretty." So I said: "Oh my God, Jerry, we didn't restage it . . . " He said, without flinching and without a second's delay: "Ad-lib it."

Could you believe, that on opening night in Washington D.C., the first time the world was going to see us, I was being asked to improvise, to ad-lib this number, and to ad-lib it in a Jerry Robbins choreographed show? I didn't go to dinner, I went directly to my dressing room and began trying to figure out what I was going to do in that space. We had no ending—the girls were told to just stand on the side by the door, and it was up to me to figure out something to do. So I got a rose (a phony rose), a mirror, a brush, and a fan and I began improvising how I would be if there was nobody there looking at me and I was making myself pretty to go out on my first date with Tony. At the end of the number there was no room for the big dance thing, so all I could think of was to jump on the bed, and like a four-year-old to twirl, twirl, twirl and then fall on my behind. And that's what I did. I finished like that and the audience went crazy. And it stopped the show!

But the most wonderful part, told to me afterwards by Arthur Laurents, was that after the show, as Oscar Hammerstein was walking up the aisle, he came over to Jerry and Arthur and Lenny and Steve, who were at the top, watching from the back row and said: "Congratulations to all of you. This is an incredible milestone in the theatre." And he raved and raved about every aspect of the show. And then, turning to Jerry, he said: "But my favorite moment in the entire show came with the spontaneity of 'I Feel Pretty.' I don't know how you did it, but you encapsulated the joy of a young woman in love. And you are to be congratulated." And Jerry said: "Thank you."

Did the other three collaborators ever try to intervene on anybody's behalf?
Only Lenny.

Do you remember another incident during the "tech" rehearsal for the first performance when Robbins leaned over Max Goberman, who was conducting, and told him he wanted certain notes left out of the score? Bernstein apparently walked out of the theatre . . .
I don't actually recall that, but you must remember that with all the lights on you can't really see what's happening in the house—it's black as pitch. And Lenny always did things quietly; he would never have made a scene. He would just have left, and

he may well have done that. There comes a point when you can't stand it anymore, when you're watching your creation being ripped apart limb from limb. There was no Court of Appeal—Jerry was the last word. And Lenny relinquished that early on. And probably rightfully so, because otherwise I'm sure Jerry would have walked out. And then we would not have had a right to the property—he owned it.

After that first performance were you aware that West Side Story *was not only going to be a "smash hit," in Bernstein's own words, but that it was probably going to be one of the classic musicals of the century?*

We always knew that, you know. We felt that if we could only get past the critics, if we could indeed be *allowed* to be this brazen group of youngsters, untried and unknown, and allow the piece to speak for itself, then we *knew* . . . You see, what it was saying touched a nerve in the American public—the headlines were screaming with the blood that was being spent [gang-warfare at that time was rife in New York]—and Lenny's music was so magnificent. Also, because of the reaction once we were in Washington and Philadelphia—the audiences there just ate it up—we knew that we only needed to get beyond the critics, those seven people whose judgment was the epitome of success or failure. And we did that.

After the Winter Garden first performance did you all wait at Sardi's for the reviews?

Oh God yes! Into the wee hours of the morning. And then Lenny would stand on a chair and read the reviews, because he read so beautifully, and we would scream and cheer. We knew we were home then. There is a photograph of Lenny taken earlier, in Washington, walking out of the National Theater in a white suit—he had just read the reviews there from the night before and he had gone to the box office and was told that we were sold out for the rest of the run—and you can see he was walking on air. And so were we, but that was only the first hurdle—the big hurdle was New York City. On opening night in New York the curtain came down and we ran to our places for the curtain call. The curtain went up and we looked at the audience, and they looked at us, and we looked at them—there was no applause—and I thought: "Dear God in heaven, they don't get it?" They were really stunned, I think. It was the first time in New York [musical theatre] history that there had been a death scene like that. The last scene was the carrying off of a body to the sound of a bell and the strains of "Somewhere" and I was following in a vale of tears . . . Then the curtain came down, and I don't think that the audience could believe that that was it. That was it? And then, as if it were choreographed, they jumped to their feet, screaming and yelling. And I burst into tears and I looked into the wings and there was Lenny. There was curtain call after curtain call, and then when it finally stopped going up and down, he walked over to me and put his arms around me, and we literally *sobbed.* So that was opening night, and then after the reviews, we knew we were home.

Arthur Laurents's book for West Side Story *remained remarkably close to the Shakespeare on which it was based* (Romeo and Juliet), *barring the ending, where Juliet dies and Maria lives. Why do you think they decided to make this particular change?*

They *were* going to "kill" Maria, she was going to die at the end. But Arthur Laurents—I guess with all the other collaborators—spoke to Richard Rodgers about it. Richard Rodgers and Oscar Hammerstein and all of these people were very friendly with each other, and you would bounce things off the people that you respected the most. So Arthur spoke to Rodgers, and they were considering how they were going to kill Maria—was she going to take the gun and then shoot herself, for example? And Rodgers said: "You know, the moment that Tony dies, Maria is dead already. Her life is over. You don't need to 'kill' her. It's sadder if she has to live on alone." And it's true. Also, had Maria died, the show would have been nothing but dead bodies everywhere you looked! The first curtain comes down on two dead bodies, and then the second would also have come down on two dead bodies, and they figured: "No, let's not do that." And I think it was a wise decision. It is a hundred times sadder for Maria, plus it gives her the final speech ["How many bullets are left?"]. You know the final speech was supposed to have been an aria?

Yes, apparently Bernstein made four or five attempts, none of them successful . . .

Yes. I never saw the aria, so obviously it never pleased the powers that be enough to make it even into rehearsal. I think that the turning point in the decision to keep the final speech was our first run-through at the Broadway Theater, for an invited audience of mainly theatre people—everybody from Lena Horne to Cheryl Crawford—and we were in just practice clothes. The piano bench was my bed, a ladder served as the balcony and the gun was just a pencil. That performance I think was the first time we got a reading on where the strengths were in the show. We were being judged by people who had been in the theatre all their lives, and that death scene, with Maria's final speech, was pin-drop quiet. And when we went off they just screamed. So the death scene was left that way, and I never got to learn that aria. That scene is the hardest in the whole work. Maria is alone and she has to deliver the message of the entire play: "Are there enough bullets for all of us? Do we either cleanse ourselves or do we all die?" It's my favorite moment in the show.

Critics wrote after West Side Story *that they had never before seen such a range of choreography in a Broadway musical. It seems many of them gave pride of place to the choreography rather than the music . . .*

Well, the score also caused a sensation. You see, there were so many new things, so many innovative firsts, so many risks that we had taken with the show. But much of the essence of the play was portrayed in dance. First there is the "Prologue" which sets the stage at the beginning, then "The Rumble" in which two people—Bernardo and Riff—are killed, then the entire "Ballet," which was our [Tony and Maria's] projection of where we wanted to go, a projection of our dreams, our Utopia. Very important parts of the play were danced. And nobody ever questioned the enormousness of Mr. Robbins's talents. I think *West Side Story* knocked people for a loop, because we were doing Robbins's hardest steps and singing Leonard Bernstein's highest notes and playing Shakespeare's hardest theme all at once, and it was wonderful. It was what Jerry expected of us, but the public really wasn't ready for it. Certainly it's the most exciting show that I've ever been in, and also I've never seen anything take the ball and run farther.

One writer has recently suggested that West Side Story *was not only a summit in terms of the musical theatre tradition, but that it destroyed Broadway, because nothing that followed it could ever come close. Would you agree?*

No. I refuse to accept that kind of pessimism. But maybe the reason it hasn't happened again is that you don't have the people with the conviction *and* the talent *and* the power of the forces we had. Do you know how many brilliant talents we had to work with? With Stephen Sondheim, Jerry Robbins, Leonard Bernstein, and Arthur Laurents—in his own right using the genius of Shakespeare—it is not surprising that it worked. And we had the freedom of breaking all the rules. I worked, along with all of the principals in the company—Chita [Rivera], and Larry and the people playing Bernardo and Riff—for a whole month without pay. For twelve to sixteen hours a day. Somebody exposed that fact after we opened and we got close to being expelled from Equity and taken out of the show because of it. You see, today's theatre and today's union are so short-sighted that they don't realize how *much* time and effort has to go into the creative process. And so to some extent the possibility of this kind of dedication has been stifled. Jerry Robbins will not do another show because of this. The unions insist on dictating certain things, hours and so on, and that's why the American public has lost an art form, and we've relinquished it to some extent to the British. And I find that very frustrating.

Bernstein also seemed to become very disenchanted with the Broadway theatre . . .

Yes. You know, I am going to be doing a salute to Lenny in March [1994] at the Metropolitan Museum of Art that will depict his contribution to the American musical theatre. In doing that research I have found the most poignant, sad things that he said. For example he said that if he had not been so crushed by the pain of *Candide* and the pain of *West Side Story*, and had chosen not to go into conducting in such a big way, who knows what might have been? You see, *West Side* was his first dream come true, at literally *changing* the face of the musical theatre, at changing the aspect of the happy ending and the score that was easily sung. With *West Side Story*, Lenny brought to musical theatre a classical discipline, he brought to the score an operatic stature, and also he brought a statement which was an indictment of society. There was always a part of Lenny which was happiest as rabbinical teacher. This show satisfied so many of his needs. If he had continued in that vein who knows what might have happened?

Taking on the post of Music Director of the New York Philharmonic in 1958 seems to have been one of the crucial decisions in Bernstein's career. Brooks Atkinson, drama critic of the New York Times *until 1960, thought that he had made the wrong choice.*[2]

Yes, you know he had always been very much under the influence of Koussevitzky, who had been like a father to him. Koussevitzky had always had a very poor view of Broadway—Broadway was just show tunes as far as he was concerned and he considered it beneath Lenny. When Koussevitzky was alive, he had always said to Lenny: "You're better than that." For me the wonderful part about Lenny was that he had the capability of writing both. It's a dilemma to have as much talent as Leonard Bernstein, because everybody's constantly pulling at you in a different direction. Felicia wanted him to have security, to have a home base for their family, and certainly the Philharmonic offered that to him. So in terms of what Koussevitzky's ambitions had been for him, and also his responsibility to his family it must have made sense for him to take the Philharmonic post at the time. Also you must remember that in *West Side Story* he had been beaten up by Robbins pretty badly. In the classical world you write what you want, and it's *your* baby, not somebody else's. *West Side Story* was Jerry Robbins's baby. So when Lenny wrote *Kaddish* or whatever, he was writing what *he* wanted. And I think his decision to take the Philharmonic was what he thought was a healthier, more responsible way of dealing with his talents and the people around him who loved him. We can't as human beings do everything and have the hindsight of saying: "If only . . . " He would not just have had to split himself up; you can't *be* the head of the Philharmonic *and* be writing a musical show, because to write for and be part of a Broadway production means twenty-four hours of your time. You know, I think perhaps Leonard Bernstein should have been triplets.

Or maybe even quadruplets? In spite of the pain Bernstein may have gone through in the collaboration on West Side Story *he probably produced his most memorable score for this work, in collaboration with Robbins.*

Yes, I think he did. They were so well suited to each other, and Jerry is certainly one of the greatest editors that ever lived. You see, Broadway is one of the most complex collaborative arts. There has to be the book and the lyricist and the choreographer and the composer and the producer and the director and so on. It worked in *West Side Story* because every link was so brilliantly constructed, but it took ten years. How many people have ten years to devote to every project? Then also we had rehearsed for about two months before opening in Washington, and we were out of town for I think about six weeks before New York. It was a time when you were allowed these refinements. More recent Broadway shows have opened with no out of town try-outs—they have sometimes begun previewing when your costumes don't fit, and you don't know what is happening with the book because they're still changing it, and the words aren't in your mouth and so on. We had the luxury of being out of town long enough to *own West Side Story.*

When Bernstein left the Philharmonic in 1969, he did so with the idea of composing. Although there were several attempts at a musical theatre collaboration[3]*, the only work that saw the light of day was* 1600 Pennsylvania Avenue, *which failed. Do you think that Bernstein had by then lost his touch?*
I heard one song from *1600*, which was very beautiful, but I never saw the book. You know, without a proper book, no matter what you write for it, you won't make it happen. The American public has to have a story they believe in, and one has to just keep writing and writing and hope that one out of the million or so is going to make it.

Lukas Foss has said that he believes West Side Story *to be Bernstein's most serious work. Would you agree?*
I think it's the most consistent and probably the most successful internationally. That's part of what you eventually have to judge a work by—does it speak to the people of Russia and China and Europe and the world, and not just the elite clique of sophisticates in New York City. One has to look at how universal a work is in its scope, and I think *West Side Story* touches everybody.

I must give you one more story about Lenny, because it was so typical of his generosity and his loyalty to the people he loved. In 1990 Larry Kert and I were going to do an act at the

Rainbow and Stars which is one of *the* places for cabaret in New York City, at the top of Rockefeller Plaza. We were going to be together for the first time in New York since *West Side Story*.

We were searching and searching for a way to begin, which is the most critical thing about any act or any show. It occurred to me that it would be wonderful if we reconstructed the way we *met* in *West Side Story*—in the dance hall. Once Larry and I had agreed about what we were going to do and which parts of our *West Side* medley we were going to do, I realized that because it was a cabaret we had only three instruments. And nobody's going to get excited by three instruments. Without an orchestral sound and the voices of the *mambo*, of the gym, it wouldn't work. So what I wanted to do was use the recording. And my brother, a lawyer, said: "You're into grand rights. And you're going to have to get permission from the only person in the world who can give it to you—Leonard Bernstein."

So I called Lenny's office and I got Sylvia Goldstein, his lawyer. I explained the whole thing to her and said that I knew I was into grand rights and that the only person capable of releasing it was Lenny. She said that she couldn't possibly dream of bothering him—this was the August before he died—that he was at Tanglewood, he wasn't feeling well, he was exhausted and very busy. I said: "Please Sylvia, we have to open soon, and this is the way I would like to do it, but if he can't, I will understand." And I said that I would be very grateful if she would get back to me.

The next morning, at the crack of dawn, she called and she said: "Oh Miss Lawrence, it's Sylvia, and the maestro says you can use *anything* you want as long as you let him see you do it." So of course I said that any night he wanted to come he would be more than welcome! Well, she called again and said that Lenny would like to come with about fifteen people. I was thrilled and I got them one long table and they arranged to come on the Saturday night, the night that we closed. So, come

Saturday I was on needles and pins, and before the show his secretary came to my dressing room and said: "We're all here, all except Lenny. He's just not well enough. He sends flowers and an apology and his love and he's thrilled for the success of the show. And he hopes you will forgive him." I said: "Of course."

And the next day, a Sunday, he died. And the following morning I had to go on *Good Morning America* and tell that story. To me it was a tribute to the fact that he hadn't changed, that he was as generous as he had always been. It was typical of the way that he had treated us every day when we were rehearsing *West Side.*

And I think that is the way that all the true giants of the theatre are. I have worked with Fred Astaire and Maurice Chevalier and Bing Crosby, and those people were always constant. They could afford to be. And Lenny was the same.

NOTES

1. Carol Lawrence has explained the workings of Oliver Smith's sets for *West Side Story* as follows: "Oliver Smith's brilliant sets were constructed to move on and off by sliding in multiple slits in the base-floor. Like something akin to a monorail car, the various set pieces appeared from various legs off stage. Then with no stagehands in sight, they magically met, attached and evolved into the bedroom or drugstore or bridal shop. The scene would be played and then in the changing light each piece would slide silently out of sight. The sets were miraculously innovative; the stagehands were never revealed and we never had to close a curtain. The audience just gasped every time the sets changed, because it was the first time it had ever been done like that."

2. During Bernstein's Philharmonic period he only produced two compositions of note—the *Chichester Psalms* and *Kaddish,* the Third Symphony. There was nothing at all for the musical theatre.

3. One projected show from the Philharmonic period, *The Skin of Our Teeth,* never saw the light of day, and *Alarums and Excursions,* a show Bernstein had later worked on with Laurents, was also never realized.

Index